Everyday
Mathematics®

The University of Chicago School Mathematics Project

STUDENT MATH JOURNAL
VOLUME 1

Mc
Graw
Hill
Education

The University of Chicago School Mathematics Project

Max Bell, Director, *Everyday Mathematics* First Edition; James McBride, Director, *Everyday Mathematics* Second Edition; Andy Isaacs, Director, *Everyday Mathematics* Third, CCSS, and Fourth Editions; Amy Dillard, Associate Director, *Everyday Mathematics* Third Edition; Rachel Malpass McCall, Associate Director, *Everyday Mathematics* CCSS and Fourth Editions; Mary Ellen Dairyko, Associate Director, *Everyday Mathematics* Fourth Edition

Authors
Max Bell, John Bretzlauf, Amy Dillard, Robert Hartfield, Andy Isaacs, Rebecca Williams Maxcy, James McBride, Kathleen Pitvorec, Peter Saecker, Robert Balfanz*, William Carroll*, Sheil Sconiers*

*First Edition Only

Fourth Edition Grade 4 Team Leader
Rebecca Williams Maxcy

Writers
Meg Schleppenbach Bates, Randee Blair, Kristin Fitzgerald, Carla LaRochelle, Sara A. Snodgrass

Open Response Team
Catherine R. Kelso, Leader; Judith S. Zawojewski, Andy Carter, John Benson

Differentiation Team
Ava Belisle-Chatterjee, Leader; Jean Capper, Martin Gartzman, Barbara Molina

Digital Development Team
Carla Agard-Strickland, Leader; John Benson, Gregory Berns-Leone, Juan Camilo Acevedo

Virtual Learning Community
Meg Schleppenbach Bates, Cheryl G. Moran, Margaret Sharkey

Technical Art
Diana Barrie, Senior Artist; Cherry Inthalangsy

UCSMP Editorial
Don Reneau, Senior Editor; Elizabeth Olin, Kristen Pasmore, Lucas Whalen

Field Test Coordination
Denise A. Porter

Field Test Teachers
Kindra Arwood, Tiffany N. Harper, Brian A. Herman, Tonya Howell, Amy Jacobs, Amy Jarrett-Clancy, Kari Lehman, Stephanie Rogers, Jenna Rose Ryan, Rachel Schrader, JoAnn Tennenbaum, Robin Zogby

Digital Field Test Teachers
Colleen Girard, Michelle Kutanovski, Gina Cipriani, Retonyar Ringold, Catherine Rollings, Julia Schacht, Christine Molina-Rebecca, Monica Diaz de Leon, Tiffany Barnes, Andrea Bonanno-Lersch, Debra Fields, Kellie Johnson, Elyse D'Andrea, Katie Fielden, Jamie Henry, Jill Parisi, Lauren Wolkhamer, Kenecia Moore, Julie Spaite, Sue White, Damaris Miles, Kelly Fitzgerald

Contributors
William Baker, John Benson, Jeanne Di Domenico, Jim Flanders, Lila Schwartz, Penny Williams

Center for Elementary Mathematics and Science Education Administration
Martin Gartzman, Executive Director; Meri B. Fohran, Jose J. Fragoso, Jr., Regina Littleton, Laurie K. Thrasher

External Reviewers
The *Everyday Mathematics* authors gratefully acknowledge the work of the many scholars and teachers who reviewed plans for this edition. All decisions regarding the content and pedagogy of *Everyday Mathematics* were made by the authors and do not necessarily reflect the views of those listed below.

Elizabeth Babcock, California Academy of Sciences; Arthur J. Baroody, University of Illinois at Urbana-Champaign and University of Denver; Dawn Berk, University of Delaware; Diane J. Briars, Pittsburgh, Pennsylvania; Kathryn B. Chval, University of Missouri–Columbia; Kathleen Cramer, University of Minnesota; Ethan Danahy, Tufts University; Tom de Boor, Grunwald Associates; Louis V. DiBello, University of Illinois at Chicago; Corey Drake, Michigan State University; David Foster, Silicon Valley Mathematics Initiative; Funda Gönülateş, Michigan State University; M. Kathleen Heid, Pennsylvania State University; Natalie Jakucyn, Glenbrook South High School, Glenview, IL; Richard G. Kron, University of Chicago; Richard Lehrer, Vanderbilt University; Susan C. Levine, University of Chicago; Lorraine M. Males, University of Nebraska-Lincoln; Dr. George Mehler, Temple University and Central Bucks School District, Pennsylvania; Kenny Huy Nguyen, North Carolina State University; Mark Oreglia, University of Chicago; Sandra Overcash, Virginia Beach City Public Schools, Virginia; Raedy M. Ping, University of Chicago; Kevin L. Polk, Aveniros LLC; Sarah R. Powell, University of Texas at Austin; Janine T. Remillard, University of Pennsylvania; John P. Smith III, Michigan State University; Mary Kay Stein, University of Pittsburgh; Dale Truding, Arlington Heights District 25, Arlington Heights, Illinois; Judith S. Zawojewski, Illinois Institute of Technology

Note
Many people have contributed to the creation of *Everyday Mathematics*. Visit http://everydaymath.uchicago.edu/authors/ for biographical sketches of *Everyday Mathematics* 4 staff and copyright pages from earlier editions.

www.everydaymath.com

Send all inquiries to:
McGraw-Hill Education
8787 Orion Place
Columbus, OH 43240

ISBN: 978-0-02-143092-5
MHID: 0-02-143092-6

Printed in the United States of America.

2 3 4 5 6 7 8 9 QVS 19 18 17 16 15

Contents

Unit 2

Unit 3

Unit 4

Activity Sheets

Much of your work in Kindergarten through third grade involved understanding mathematics and its uses. You learned to solve number stories and use arithmetic, including basic facts in addition, subtraction, multiplication, and division.

Fourth Grade Everyday Mathematics builds on this experience, introducing more complex mathematical concepts and ways to use math in your everyday life.

Here are some things you will be asked to do in *Fourth Grade Everyday Mathematics*:

- Increase your "number sense," "measurement sense," and estimation skills.
- Extend your skills with addition, subtraction, multiplication, and division of whole numbers.
- Find equivalent fractions.
- Use your understanding of whole number operations to add and subtract fractions and multiply a fraction by a whole number.
- Learn about using letters to stand for unknown numbers and other beginning algebra concepts so you can write algebraic equations.
- Increase your understanding of measurement in both the metric and U.S. customary systems.
- Expand your ability to collect, organize, and interpret data.
- Develop your skills with 2-dimensional geometry, working with angles, classifying polygons, and investigating symmetry.

We hope that you find the activities fun and that you begin to see the beauty in mathematics. More importantly, we hope you become better at using mathematics to solve interesting problems in your own life.

Finding the Values of Digits

Use a place-value tool to solve.

SRB
78-79

1 In 9,027, what is the value of the . . .

9? _____

0? _____

2? _____

2 In 82,075, what is the value of the . . .

8? _____

2? _____

5? _____

3 **a.** The value of 1 in 10 is _____ times as large as the digit 1.

b. The value of 5 in 500 is _____ times as large as the value of the 5 in 50.

4 **a.** The value of 4 in 400 is _____ times as large as the value of 4 in 40.

b. The value of 9 in 9,000 is

_____ times as large as the value of 9 in 900.

5 **a.** Write a number in which the digit 7 is worth 70. _____

b. Write a new number in which the digit 7 is worth 10 times

as much as the number you just wrote. _____

6 Write the number that has . . .

3 in the ten-thousands place
4 in the ones place
2 in the thousands place
6 in the hundreds place
0 in the remaining place

____ ____, ____ ____ ____

7 Write the number that has . . .

8 in the hundreds place
6 in the hundred-thousands place
3 in the tens place
1 in the ten-thousands place
7 in the ones place
0 in the remaining place

____ ____ ____, ____ ____ ____

Try This

8 Write the number name for 5,622,463. _____

_____

Math Boxes

1 Add mentally.

a. 8 + 5 = _____

b. 80 + 50 = _____

c. 7 + 7 = _____

d. 70 + 70 = _____

e. 6 + 9 = _____

f. 60 + 90 = _____

2 Fill in the missing numbers and state the rule.

a. 2, 4, 6, _____*8*_____, _____, _____

Rule: _*+ 2*_

b. 65, 60, 55, _____, _____, _____

Rule: _____

c. 109, 95, 81, _____, _____, _____

Rule: _____

3 What is the value of the digit 5 in **5**60?

500

What is the value of the digit 7 in the numbers below? Use a place-value tool.

a. 4**7**5 _____

b. **7**,058 _____

c. 9,9**4**7 _____

d. 68,0**7**1 _____

SRB
78-79

4 What number is halfway between 300 and 600? Fill in the circle next to the best answer.

(A) 550

(B) 450

(C) 525

(D) 400

5 **Writing/Reasoning** Round 475 to the nearest ten. _____ Explain how you rounded the number.

_____

_____

_____

SRB
85-87

3

Hockey Game Attendance

Write < or >.

1 508 _____ 588

2 2,675 _____ 3,675

3 1,833 _____ 1,933

4 57,883 _____ 56,838

SRB
81

This table shows the total attendance for the 2013–2014 regular season for ten National Hockey League (NHL) teams.

NHL Team	2013–2014 Regular Season Total Attendance
Buffalo	761,767
Carolina	634,832
Chicago	927,545
Detroit	908,131
Minnesota	758,729
Montreal	872,193
Philadelphia	813,411
Pittsburgh	763,344
Tampa Bay	763,096
Vancouver	810,594

Source: www.espn.go.com/nhl/attendance

5 Which team had the largest

total attendance? _____

Which team had the smallest

total attendance? _____

6 Compare the attendance for Philadelphia and Minnesota.

  a. Which team had the greater

  attendance? _____

  How do you know?

  _____

  _____

  b. Write a comparison number sentence. _____

7 Compare the attendance for Vancouver and Montreal.

  a. Which team had the greater attendance? _____

  How do you know? _____

  _____

  _____

  _____

  b. Write a comparison number sentence. _____

8 Compare the attendance for Pittsburgh and Tampa Bay.

   a. Which team had the smaller attendance? _____ How do you know?

   _____

   _____

   _____

   _____

   b. Explain how you compared the numbers. _____

   _____

   _____

   c. Write a comparison number sentence. _____

9 List the teams in order from greatest to least attendance.

NHL Team	2013–2014 Regular Season Total Attendance

Math Boxes

① Subtract mentally.

a. 7 − 4 = _____

b. 70 − 40 = _____

c. 14 − 5 = _____

d. 140 − 50 = _____

e. _____ = 16 − 7

f. _____ = 160 − 70

② **a.** Write the largest number you can make with the digits 5, 2, 3, 0, 6, 0. Use each digit only once.

_____

b. Use the same digits to write the smallest number you can make. Do not start with 0.

_____

SRB
78-79

③ Write >, <, or = to make each number sentence true.

a. 45 + 10 ____ 55 − 10

b. 387 ____ 300 + 87

c. 4 thousand ____ 4,502

d. 8,000 + 3,000 + 400 ____ 8,400

SRB
81

④ What is another name for 40? Circle the best answer.

A. 5 * 80

B. 20 + 20 + 20

C. 8 * 5

D. 50 − 20

⑤ Add.

a.   4 3 2
     + 7 3 5
     ―――――

b.   9 3 8
     + 3 6 6
     ―――――

⑥ Write the number forty-seven thousand, three hundred ninety-two in standard form.

_____

What is the value of the 4?

_____

SRB
80

6

Math Boxes

Rounding

Complete the number lines, plot the numbers on the number lines, and then use them to help you round the numbers.

SRB 85-87

1 Round 4,698 to the nearest thousand. Rounded number _____

<------------+-------------------+-------------------+------------->

_____   _____   _____
lower number   halfway number   higher number

2 Round 5,778 to the nearest ten. Rounded number _____

<------------+-------------------+-------------------+------------->

5,770   _____   _____
lower number   halfway number   higher number

3 Round 2,304 to the nearest ten. Rounded number _____

<------------+-------------------+-------------------+------------->

_____   _____   _____
lower number   halfway number   higher number

4 Round to the nearest hundred-thousand.

   a. 271,009 _____   **b.** 809,050 _____   **c.** 282,582 _____

   d. 450,076 _____   **e.** 723,670 _____   **f.** 596,207 _____

5 Round to the nearest ten-thousand.

   a. 48,300 _____   **b.** 97,042 _____   **c.** 19,203 _____

   d. 170,500 _____   **e.** 124,897 _____   **f.** 209,800 _____

6 Round to the nearest thousand.

   a. 1,200 _____   **b.** 7,500 _____   **c.** 2,582 _____

   d. 88,888 _____   **e.** 13,004 _____   **f.** 19,652 _____

Try This

7 When Claude rounded 49,540 to the nearest thousand, his result was 41,000.
   Is Claude correct? If not, what should he have done?

   _____

   _____

Math Boxes

Math Boxes

1 Add mentally.

a. 3 + 3 = _____

b. 30 + 30 = _____

c. 5 + 6 = _____

d. 50 + 60 = _____

e. 4 + 7 = _____

f. 40 + 70 = _____

2 Fill in the missing numbers and state the rule.

a. 4, 8, 12, 16, _____, _____, _____

Rule: _____

b. 33, 30, 27, _____, _____, _____

Rule: _____

c. _____, _____, _____, 106, 141, 176

Rule: _____

3 Use a place-value tool to help you determine the value of the digit 8 in the numbers below.

a. 7,5**8**4 _____

b. 3**8**,067 _____

c. 49,**8**41 _____

d. **8**20,731 _____

e. 391,46**8** _____

SRB
78-79

4 Complete.

a. Is 63 closer to 60 or 70? _____

b. What number is halfway between 80 and 90? _____

c. Is 572 closer to 500 or 600? _____

d. What number is halfway between 500 and 800? _____

5 **Writing/Reasoning** Explain how the value of the digit 5 in the number 5,555 changes as you move from right to left.

_____

_____

_____

SRB
78-79

Working with Populations

Use the information in the table to solve the problems.

SRB
78-79,
81

Populations of Cities		
City	2000 Census	2010 Census
New York City	8,008,278	8,175,133
Los Angeles	3,694,820	3,857,799
Houston	1,953,631	2,160,821
Philadelphia	1,517,550	1,547,607
San Diego	1,223,400	1,338,348
Boston	589,141	617,594
Milwaukee	596,974	594,833
Albuquerque	448,607	545,852
St. Paul	287,151	285,068
Norman	95,694	110,925

Source: U.S. Census

1　Name two cities that have a 2010 population in the hundred-thousands.

_____

2　Name two cities that have a 2010 population in the millions.

_____

3　Boston's population in 2010 was 617,594. What is the value of the digit . . .

1? _____　　7? _____　　6? _____

4　Philadelphia's population in 2010 was 1,547,607. What is the value of the digit . . .

4? _____　　1? _____　　5? _____

5　Round the 2010 population of Houston to the nearest million. _____

6　Did Boston's population increase or decrease from 2000 to 2010? _____

7　Record the population for Norman in 2000 and 2010. Use <, >, or = to compare.

_____

8　Which cities had populations that decreased from 2000 to 2010?

_____

9

Math Boxes

Math Boxes

1   Subtract mentally.

    a. $15 - 7 =$ _____

    b. $150 - 70 =$ _____

    c. $13 - 8 =$ _____

    d. $130 - 80 =$ _____

    e. $18 - 9 =$ _____

    f. $180 - 90 =$ _____

2   What is the largest number you can make with the digits 3, 0, 3, 8, and 0? Fill in the circle next to the best answer.

    Ⓐ 83,003

    Ⓑ 83,030

    Ⓒ 83,300

    Ⓓ 80,033

SRB 78-79

3   Write >, <, or = to make these number sentences true.

    a. $952 - 52$ _____ $850 + 50$

    b. 3 thousand _____ 3,001

    c. $2,100$ _____ $2,000 - 100$

    d. 9 [10,000s] _____ 90,000

SRB 81

4   Cross out the 3 names that do not belong in the name-collection box. Label the box with the correct number.

$25 - 13$
$20 - 7$
$6 * 2$
$4 * 3$
$40 - 23$
$7 * 3$

5   Add.

    a. $4,563 + 8,712 =$ _____

    b. $3,656 + 7,997 =$ _____

6   Write 842,176 in words. Use a Place-Value Flip Book or chart to help you.

_____

_____

_____

_____

_____

SRB 79

Estimating Sums and Differences

Read the number stories. Choose an appropriate estimation strategy.

SRB
82-89

1 Workers in a toy factory make stuffed bears. The goal is to make at least 1,200 bears every 3 hours. In 1 hour they made 447 bears. In the next hour they made 453 bears. In the third hour they made 458 bears.

   a. Did they meet the 3-hour goal? _____

   b. How did you get your answer? _____

      _____

   c. Why did you choose this estimation strategy? _____

      _____

      _____

      _____

      _____

2 Stella is saving up to buy a bicycle that costs $250. She has saved $96 so far. Her aunt gave her another $48.

   a. About how much more money does Stella need? About _____

   b. How did you get your answer? _____

      _____

   c. Why did you choose this estimation strategy? _____

      _____

      _____

      _____

Math Boxes

1 Subtract mentally.

a. $15 - 8 =$ _____

b. $150 - 80 =$ _____

c. $13 - 7 =$ _____

d. $130 - 70 =$ _____

e. $14 - 6 =$ _____

f. $140 - 60 =$ _____

2 Use a Place-Value Flip Book or chart. In the number 742,318, the 2 stands for 2,000.

a. The 1 stands for _____.

b. The 8 stands for _____.

c. The 4 stands for _____.

d. The 3 stands for _____.

e. The 7 stands for _____.

SRB
78-79

3 Put these numbers in order from least to greatest.

32,000　　　3,200

23,000　　　2,300

_____

_____

_____

_____

SRB
81

4 a. Round 6,245 to the nearest . . .

thousand _____

ten _____

b. Round 591,340 to the nearest . . .

hundred-thousand _____

ten-thousand _____

SRB
85-87

5 **Writing/Reasoning** What steps do you follow when you round a number?

_____

_____

_____

_____

_____

_____

_____

SRB
85-87

World's Tallest Buildings

Math Message

SRB
82-84

1. Terrell is a competitive stair climber. He competes by climbing stairs as fast as he can. This year he wants to climb 400 stories. The Burj Khalifa in the United Arab Emirates is the world's tallest building with 163 stories. The Petronas Towers, the tallest building in Malaysia, has 88 stories. The Zifeng Tower in China has 89 stories. If Terrell climbs these three buildings, how many more stories will he need to climb to reach his goal of 400?

   Estimate: _____ stories

   Answer: _____ stories

   Number model with answer: _____

Estimate and then solve.

2. Terrell climbed four buildings in the United States. In Chicago, he climbed the Willis Tower (108 stories) and the John Hancock Center (100 stories). In New York City, he climbed the Empire State Building (102 stories) and the Bank of America Tower (54 stories). How many stories did he climb in all?

   Estimate: _____ stories

   Answer: _____ stories

   Number model with answer: _____

   Does your answer make sense? _____ How do you know? _____

   _____

   _____

3. Terrell climbed four buildings for a total of 472 stories. He climbed the Burj Khalifa (163 stories), the Abraj Al Bait in Saudia Arabia (120 stories), the Taipei 101 in Taiwan (101 stories), and another tall building. How many stories tall is the last building he climbed?

   Estimate: _____ stories

   Answer: _____ stories

   Number model with answer: _____

   Does your answer make sense? _____ How do you know? _____

   _____

   _____

13

World's Tallest Buildings (continued)

4 Two of China's tallest buildings are the Shanghai World Financial Center (101 stories) and the Zifeng Tower (89 stories). Two of the tallest buildings in the United States are the Willis Tower (108 stories) and the Trump Tower (98 stories). How many more stories do the two U.S. buildings have than the two buildings in China?

Estimate: _____ stories

Answer: _____ stories

Number model with answer: _____

Does your answer make sense? _____ How do you know?

_____

_____

Try This

5 The International Commerce Centre in Hong Kong has 118 stories. The Jin Mao Tower in Shanghai and Two International Finance Centre in Hong Kong have the same number of stories. If the 3 buildings together have 294 stories, how many stories do the Jin Mao Tower and Two International Finance Centre each have?

What is the unknown? _____

_____

_____

Estimate: _____ stories each

Answer: _____ stories each

Number model with answer: _____

Does your answer make sense? _____ How do you know?

_____

_____

14

Math Boxes

① Subtract mentally.

a. $11 - 2 =$ _____

b. $110 - 20 =$ _____

c. _____ $= 12 - 4$

d. $120 - 40 =$ _____

e. _____ $= 9 - 7$

f. $90 - 70 =$ _____

② Which expanded form shows 3,458?
Fill in the circle next to ALL that apply.

◯ a. 3 [1000s] + 4 [100s] + 5 [10s]
      + 8 [1s]

◯ b. 300 + 400 + 50 + 8

◯ c. 3,000 + 45 + 8

◯ d. 3,000 + 400 + 50 + 8

SRB
80

③ Subtract. Show your work.

a. 6 6 4
   − 3 9 2
   _____

b. 4 2 4
   − 2 9 6
   _____

SRB
94-101

④ Make an estimate. Write a number model
to show your strategy.

a. 3,389 + 2,712

_____

b. 3,452 − 1,147

_____

SRB
82-89

⑤ Use a place-value tool. A number has . . .

6 in the hundreds place

2 in the tens place

8 in the hundred-thousands place

5 in the ones place

3 in the thousands place

4 in the ten-thousands place

Write the number.

___ ___ ___, ___ ___ ___

SRB
78-79

⑥ Write >, <, or = to make each number
sentence true.

a. 14 ____ 26

b. 3,003 ____ 3,300

c. 12 + 12 ____ 24

d. 200 − 50 ____ 100

e. 30 + 30 ____ 50 + 10

SRB
81

15

U.S. Traditional Addition

SRB
92-93

Make an estimate. Write a number model to show your thinking. Try to solve each problem using U.S. traditional addition. Compare your answer with your estimate to see whether your answer makes sense.

1
```
   4 9
 + 3 3
```

Estimate: _____

2 94 + 47 = _____

Estimate: _____

3
```
   3 7 2
 + 4 8 9
```

Estimate: _____

4
```
   4 6 2
 + 9 4 9
```

Estimate: _____

5
```
   5 3 8
 + 9 2 8
```

Estimate: _____

Try This

6 4,674 + 6,053 = _____

Estimate: _____

7 There are 279 boys and 347 girls at a school assembly.
How many students are at the assembly?

Estimate: _____ students

Answer: _____ students

Comparing Addition Strategies

Estimate 356 + 498. Write a number model to show your thinking. Then solve using partial-sums addition, column addition, and U.S. traditional addition.

SRB
90-93

Estimate: _____

Partial-Sums Addition

Column Addition

U.S. Traditional Addition

Which method do you prefer? Why?

_____

_____

_____

_____

Math Boxes

Math Boxes

① Subtract mentally.

   a. $17 - 7 = $ _____

   b. $170 - 70 = $ _____

   c. $15 - 9 = $ _____

   d. _____ $= 150 - 90$

   e. $16 - 8 = $ _____

   f. _____ $= 160 - 80$

② In the number 30,516, what does the 3 stand for? Circle ALL that apply.

   A.   3,000

   B.   3 [10,000s]

   C.   30,000

   D.   300,000

SRB
78-79

③ Put these numbers in order from least to greatest.

   46,000   64,000

   4,600   4,006

   _____

   _____

   _____

   _____

SRB
81

④ **a.** Round 81,886 to the nearest . . .

   thousand _____

   ten-thousand _____

   b. Round 245,197 to the nearest . . .

   hundred-thousand _____

   ten-thousand _____

SRB
85-87

⑤ **Writing/Reasoning** What do you need to consider when you order the whole numbers in Problem 3?

_____

_____

_____

_____

_____

SRB
81

Grouping by 25s, 5s, and 1s

At Barbara's Bagel Bakery, Bob packs bagels into boxes that hold 25 bagels, 5 bagels, or 1 bagel.

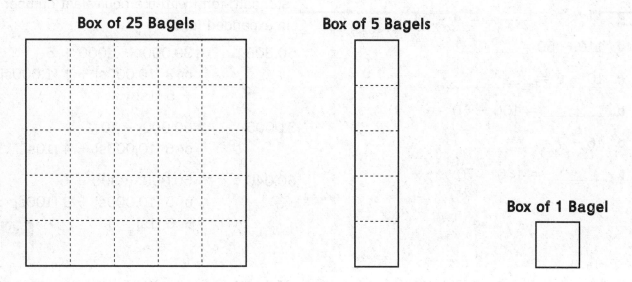

Box of 25 Bagels   **Box of 5 Bagels**

Box of 1 Bagel

Bagels come to Bob on a tray. Bob always fills the largest box possible and makes sure each box is full. For each tray of bagels, how many boxes of each size does Bob fill? Complete the table.

Number of Bagels on the Tray	Boxes of 25 Bagels	Boxes of 5 Bagels	Boxes of 1 Bagel
27			
10			
53			
9			

Explain to your partner how you figured out your answers.

Math Boxes

Math Boxes

---

1 Subtract mentally.

a. $11 - 5 =$ _____

b. $110 - 50 =$ _____

c. $10 - 6 =$ _____

d. _____ $= 100 - 60$

e. $16 - 7 =$ _____

f. _____ $= 160 - 70$

---

2 Draw a line connecting each number in standard form with the equivalent number in expanded form.

50,306 $30,000 + 1,000 + 5$
                or 3 [10,000s] + 1 [1,000s]
                + 5 [1s]

31,005 $60,000 + 40$
                or 6 [10,000s] + 4 [10s]

60,040 $50,000 + 300 + 6$
                or 5 [10,000s] + 3 [100s]
                + 6 [1s]

SRB 80

---

3 Subtract.

a.   8 4 2   b.   7, 4 6 8
    − 5 2 1   − 3, 3 9 4

SRB 94-101

---

4 Make an estimate. Write a number model to show your strategy.

a. $1,459 + 291$

_____

b. $681 - 346$

_____

SRB 82-89

---

5 Use a Place-Value Flip Book or chart. Write the number that has . . .

1 in the ones place
8 in the thousands place
9 in the ten-thousands place
0 in the tens place
6 in the hundred-thousands place
5 in the hundreds place

____ ____ ____, ____ ____ ____

SRB 78-79

---

6 Write >, <, or = to make each number sentence true.

a. $16 + 11$ ____ 47

b. 206 ____ 602

c. $150 - 50$ ____ 100

d. $62 + 10 + 10$ ____ $62 - 10 - 10$

e. $423,726$ ____ $413,999$

SRB 81

U.S. Traditional Subtraction

Make an estimate. Write a number model to show your thinking. Try to solve each problem using U.S. traditional subtraction. Compare your answer with your estimate to see whether your answer makes sense.

SRB 82-84, 100-101

①
```
   5 8
 - 3 9
```

Estimate: _____

② 94 − 56 = _____

Estimate: _____

③
```
   6 0 0
 - 3 7 9
```

Estimate: _____

④
```
   8 3 6
 - 7 8 2
```

Estimate: _____

⑤
```
  5, 1 7 2
 -    2 3 4
```

Estimate: _____

Try This

⑥
```
  8, 0 0 4
 - 3, 5 0 6
```

Estimate: _____

⑦ The drive to Yuri's grandmother's house is 642 miles. Yuri's family has driven 484 miles so far. How many miles do they have left to drive?

Estimate: _____ miles

Answer: _____ miles

21

Comparing Subtraction Strategies

Estimate 825 − 478. Write a number model to show your thinking. Then solve using each of the three subtraction methods: counting-up, trade-first, and U.S. traditional subtraction.

Estimate: _____

SRB
94-101

Counting-Up Subtraction

Trade-First Subtraction

U.S. Traditional Subtraction

Which method do you prefer? Why?

_____

_____

_____

_____

_____

Math Boxes

1 Subtract.

a.
```
    8 7 6
  − 4 4 1
```

b.
```
    6 5 2
  − 5 3 8
```

SRB 94-101

2 Jen's goal is to hike about 200 miles on Wisconsin's state trails. If she hikes 74 miles on Tuscobia, 49 miles on Glacial Drumlin, and 24 miles on Great River, how many more miles will she need to hike to reach her goal? Show your work.

Answer: _____ miles

SRB 83, 92-93, 100

3 Write >, <, or = to make each number sentence true.

a. 67 − 10 _____ 57

b. 11 thousand _____ 11,300

c. 400 + 40 + 10 _____ 400 + 50

d. 5,000 − 26 _____ 5 thousand

SRB 81

4 Solve using U.S. traditional addition.

a.
```
    8 7
  + 9 6
```

b.
```
    2 3 9
  + 2 8 5
```

SRB 92-93

5 **Writing/Reasoning** Describe how you used U.S. traditional addition to solve Problem 4a.

_____

_____

_____

_____

_____

_____

SRB 92-93

23

Units of Length

Measurement Scales

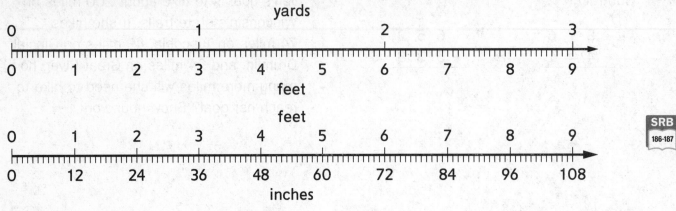

SRB
186-187

Convert.

1

Feet	Inches
1	12
2	
3	
5	

2

Yards	Feet
1	3
2	
4	
5	

3

Feet	Inches
7	
9	
	120
20	

4

Yards	Feet
7	
	27
10	
20	

Solve the problem. Complete the measurement scale to convert.

5   An Andean condor is about 4 feet tall. A raven is about 2 feet tall.
What is the combined height of the two birds in inches? _____ inches

feet

0   1   2   3   4   5   6

0   12

_____   _____   _____   _____   _____

inches

24

Units of Length (continued)

6 The saltwater crocodile can grow to be 7 yards long. The manatee and the American alligator can each grow to be 5 yards long. What is the combined length of the three animals in feet?

Answer: _____ feet

7 The giraffe is the tallest land animal in the world. It can be up to 19 feet tall. The height of the tallest giraffe combined with the height of an African elephant is 35 feet. What is the height of the elephant in inches?

Answer: _____ inches

8 On average, a blue whale is 28 yards in length. A North Pacific right whale is 17 yards in length. What is the difference in length between these two whales in feet?

Answer: _____ feet

Try This

9 The reticulated python is the longest snake in the world. It can measure up to 33 feet. The Barbados thread snake is the smallest known species of snake. It averages slightly under 4 inches in length. Estimate the difference between the length of the Barbados thread snake and the reticulated python.

Estimate: _____ feet

Explain how you got your answer. _____

_____

_____

_____

_____

25

Math Boxes
Preview for Unit 2

1 Make an array for

a. 4 * 4 _____

b. 3 * 6

SRB
53

2 Multiply mentally.

a. 2 * 1 = _____

b. _____ = 5 * 0

c. _____ = 5 * 2

d. 5 * 4 = _____

e. 3 * 10 = _____

3 Fill in the missing numbers.

a. _____, _____, _____, 50, 55, 60

Rule: _____

b. _____, _____, _____, 22, 24,

_____

Rule: _____

c. _____, _____, 42, _____,

_____, 60

Rule: _____

4 Gail was counting geese as they flew over her field. The first group contained 27 geese. The second group had 7 more geese than the first group.

Which number model(s) show(s) how many geese are in the second group? Select ALL that apply.

☐ 27 − 7 = g

☐ 27 + 7 = g

☐ 27 = g + 7

☐ g = 27 + 7

SRB
37, 47

5 Find the area of the rectangle.

3 cm

4 cm

Area: _____ square cm

SRB
202-204

6 Josephine makes and sells pottery vases. If she charges $9 per vase and she sells 6 vases one week and 7 vases the next, how much will she earn? Show your work.

Estimate: $ _____

Answer: $ _____

SRB
83

Points, Line Segments, Lines, and Rays

Use a straightedge to draw the following:

1 **a.** Draw and label line segment RT ($\overline{RT}$).

   b. What is another name for $\overline{RT}$? _____

2 **a.** Draw and label line BN ($\overleftrightarrow{BN}$). Draw and label a point T on it.

   b. What are two other names for $\overleftrightarrow{BN}$? _____

3 **a.** Draw and label ray SL ($\overrightarrow{SL}$). Draw and label a point R on it.

   b. What is another name for $\overrightarrow{SL}$? _____

   c. Under ray SL ($\overrightarrow{SL}$), draw and label ray XM ($\overrightarrow{XM}$) so it is parallel to ray SL ($\overrightarrow{SL}$).

4 A •   • B

  D •   • C

   a. Using points A, B, C, and D, create a shape that has 2 pairs of parallel line segments.

   b. Name the parallel line segments. _____

_____

Math Boxes

1 Kenji is following a 22-week training schedule to prepare for a marathon. The last four weeks call for these weekly running totals: 58 km, 45 km, 37 km, and 29 km. How far will Kenji run in the last four weeks of training? Show your work.

Estimate: _____

Answer: _____ km

SRB
92-93

2 Draw and label line QR ($\overleftrightarrow{QR}$). Draw point S on it.

What are two other names for line QR?

_____

SRB
226-227

3 Write the number in standard form.

a. 9 [100s] + 3 [1s] = _____

b. 5 [1,000s] + 4 [10s] = _____

c. 4 [10,000s] + 5 [1,000s] + 6 [100s]

+ 9 [1s] = _____

d. 2 [100,000s] + 6 [1,000s] + 7 [100s]

+ 4 [1s] = _____

SRB
80

4 Name as many rays as you can in the figure below.

_____

_____

_____

SRB
226-227

5 **Writing/Reasoning** Explain how you used your estimate to see if your answer to Problem 1 was reasonable.

_____

_____

SRB
83

28

Angles

1 Draw ∠BAC. What is another name

for ∠BAC? _____

C •

2 What is the vertex of ∠BAC? Point _____

A •   •
                       B

3 **a.** What type of angle is angle BAC in Problem 1? _____

b. How do you know? _____

---

4 Feng said the name of this angle is ∠SRT. Is he right? Explain.

_____

_____

R

S

T

5 Use the points shown on the grid below and a straightedge to draw triangle CDE.

a. What type of angle is angle DCE? _____

b. What type of triangle is triangle CDE? _____

c. Name the perpendicular line segments. _____

C •   D •

E •

29

Math Boxes

1 Subtract using U.S. traditional subtraction.

a.
```
    3 3 4
  − 2 3 8
```

b.
```
    8 8 1
  − 4 3 6
```

SRB
100-101

2 Eric's car travels about 256 miles on a full tank of gas. With the gas tank full, Eric drove 66 miles to visit cousins. Then he drove 78 miles to visit Grandma. How many more miles can Eric drive before he runs out of gas? Show your work.

Answer: _____ miles

SRB
83, 92–
93, 100

3 Write >, <, or = to make each number sentence true.

a. 55,699 _____ 45,609

b. 67,749 _____ 66,749

c. 858,193 _____ 808,192

d. 2 thousand _____ 20 hundred

e. 208,775 _____ 2 million

SRB
83

4 Add using U.S. traditional addition.

a.
```
    3 5 4
  + 5 8 9
```

b.
```
    8 0 9
  + 6 9 3
```

SRB
92-93

5 **Writing/Reasoning** Explain how you write Problem 4b in expanded form.

_____

_____

_____

SRB
80

Math Boxes

1 In Kendra's city, most blocks are about 328 feet long. If Kendra runs 3 blocks before she rests, about how many feet will she have run?

- (A) 656 feet

- (B) 1,084 feet

- (C) 3,328 feet

- (D) 984 feet

- (E) 331 feet

SRB
92-93

2 Draw and label line AB.

Draw point C on it.

What are two other names for line AB?

_____

SRB
226-227

3 **a.** 600,000 + 5,000 + 700 + 8 is the expanded form for what number?

_____ _____ _____, _____ _____ _____

b. 3,000,000 + 200,000 + 6,000 + 40 + 7 is the expanded form for what number?

_____, _____ _____ _____, _____ _____ _____

SRB
80

4 Name as many line segments as you can in the figure below.

Q •――――――• R

T •――――――• S

_____

_____

SRB
226-227

5 **Writing/Reasoning** Describe how you could use U.S. traditional addition to solve Problem 1.

_____

_____

_____

_____

SRB
92-93

31

Finding the Perimeter

Math Message

2 feet   Find the perimeter of the square. _____ feet

How did you find the perimeter? _____

_____

_____

How many inches is that? _____ inches. Explain how you converted from feet to inches.

_____

_____

1

width 9 inches

length
15 inches   What is the perimeter? _____ inches

Write an equation for the perimeter of the rectangle.

Equation: _____ inches

2 Measure the lengths and widths of your journal and 2 different rectangular objects
in your classroom. Measure to the nearest inch. Record the measurements.
Use the measurements to find the perimeter of each object.

Object Measured	Length	Width	Formula	Perimeter
Math Journal				

Perimeter Formulas for Rectangles:

$$p = l + l + w + w \qquad p = 2l + 2w \qquad p = 2 * (l + w)$$

3 Use a formula to find the perimeters of the rectangles.

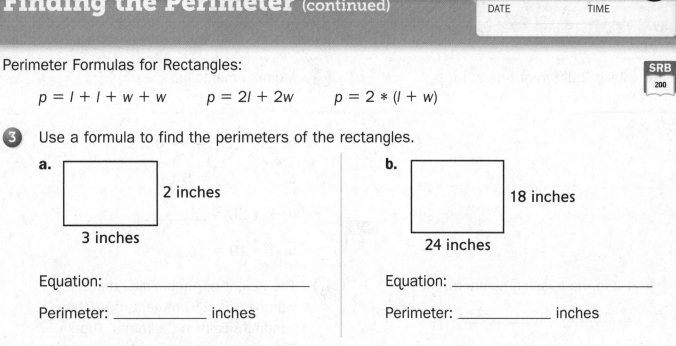

a.

2 inches

3 inches

Equation: _____

Perimeter: _____ inches

b.

18 inches

24 inches

Equation: _____

Perimeter: _____ inches

4 Jerry wants to build a rectangular vegetable garden with a fence around it. He wants the garden to be 8 feet long and 4 feet wide. Sketch his garden. Find the perimeter. Show your work.

Perimeter: _____ feet

5 Moonja's parents are building a deck. A diagram is to the right. They want the perimeter to be 44 feet. One side must be 12 feet long. What is the measurement of the width?

?

12 feet

Width: _____ feet

6 Draw a rectangle with a perimeter of 16 centimeters. Label the lengths of the sides.

Math Boxes

1 Make 2 different arrays for 6.

SRB
53

2 Multiply mentally.

a. $8 * 1 =$ _____

b. _____ $= 9 * 0$

c. _____ $= 5 * 6$

d. $5 * 50 =$ _____

e. $7 * 10 =$ _____

3 Fill in the missing numbers.

a. 16, 20, 24, _____,

_____, _____

Rule: _____

b. 15, _____, 21, _____,

27, _____

Rule: _____

c. _____, _____, 30,

_____, _____, 60

Rule: _____

4 The height of Angel Falls, the tallest waterfall, is 240 meters more than Yosemite Falls in California. Tugela Falls, the second highest waterfall, is 31 meters shorter than Angel Falls. If Yosemite Falls is 739 meters high, how high is Tugela Falls?

A.   979 meters

B.   948 meters

C.   1,010 meters

D.   468 meters

SRB
47, 83

5 Shade tiles in the grid to make an array for 6 × 6. How many tiles did you shade?

_____

SRB
202-204

6 On average, a painter can cover about 300 square feet with 1 gallon of paint. If the painter has 6 gallons of paint in his van and 5 gallons in his store, about how many square feet can he cover?

Estimate: _____ square feet

Answer: _____ square feet

SRB
83

Square Numbers

A **square array** is a special rectangular array that has the same number of rows and columns. A square array represents a whole number called a **square number.**

The first four square numbers and their arrays are shown below.

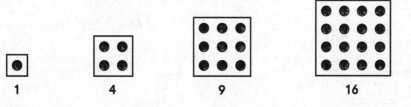

1   4   9   16

1. Draw square arrays to represent the next two square numbers after 16.
   Write each square number.

   a.   **b.**

   Square number: _____   Square number: _____

2. List all of the square numbers through 50. Use counters or draw arrays, if needed.

   _____

3. How can you find the tenth square number without drawing the array?

   _____

   _____

4. List the first 10 square numbers.

   _____

   Describe any patterns you see.

   _____

   _____

35

Math Boxes

1 Solve.

a. $5 * 7 =$ _____

b. $2 *$ _____ $= 12$

c. _____ $* 7 = 49$

d. $9 *$ _____ $= 45$

e. $8 * 4 =$ _____

2 Draw $\angle MRT$.

T •

M •

• R

What is another name for $\angle MRT$?

_____

SRB
228

3 Write >, <, or = to make each number sentence true.

a. 46,609 _____ 45,699

b. 67,749 _____ 66,749

c. 2 thousand _____ 2,000

d. 108,755 _____ 1 million

e. 1,000,000 _____ 858,192

SRB
81

4 Use U.S. traditional subtraction.

a. $\begin{array}{r} 1,247 \\ -\ 156 \\ \hline \end{array}$ b. $\begin{array}{r} 3,531 \\ -\ 2,246 \\ \hline \end{array}$

SRB
100-101

5 Keanu has 10 feet of model train track. His sister has 14 feet, and his brother has 6 feet. How many inches of track do they have all together?

Answer: _____ inches

SRB
186-187

6 Circle the right angles.

SRB
229

36

Finding the Area of Rectangles

Draw the rectangle on the grid. Then fill in the blanks.

1 Draw a rectangle with a length of 5 and width of 8.

Equation: _____

Area = _____ square units

2 Draw a 6-by-6 rectangle.

Equation: _____

Area = _____ square units

Find the area of each rectangle below.

3

3 cm

5 cm

Length: _____ cm

Width: _____ cm

Equation: _____

Area = _____ square centimeters

4

4 cm

6 cm

Length: _____ cm

Width: _____ cm

Equation: _____

Area = _____ square centimeters

5 Mrs. Mobasseri wants to cover a board with fabric. The board is a rectangle that is 9 feet long and 5 feet high. How many square feet of fabric will Mrs. Mobasseri need?

Equation: _____

Area = _____ square feet

37

Math Boxes

1 **a.** The dresser is 3 feet tall. What is the height in inches?

_____ inches

b. The sidewalk is 15 yards long. What is the length in feet?

_____ feet

SRB
186-187

2 Regulation football fields are rectangles that measure 360 feet long and 160 feet wide. If Jamie walks the perimeter of the football field, about how far will she go?

Number model:

_____

_____

Answer: _____ feet

SRB
92-93,
200-201

3 Fill in the missing digits.

```
   5,  3  8  □
+  2,  □  7  7
_____
   7,  8  6  3
```

SRB
92-93

4 Draw lines to match each word to the correct pair or pairs of line segments.

perpendicular

parallel

intersecting

SRB
230-231

5 **Writing/Reasoning** How did you find the missing digits in Problem 3?

_____

_____

_____

_____

_____

SRB
92-93

Finding Factor Pairs

1. Write equations to help you find all the factor pairs of each number below.
   Use dot arrays, if needed.

SRB
53

Number	Equations with 2 Factors	Factor Pairs
20		
16		
13		
27		
32		

2. How do you know when you have found all the factor pairs for a number?

_____

_____

39

Math Boxes

1 Multiply mentally.

a. $5 * 8 =$ _____

b. $2 *$ _____ $= 16$

c. _____ $* 7 = 21$

d. $9 *$ _____ $= 81$

e. $8 * 3 =$ _____

2 Draw ∠TIF. What is the vertex of ∠TIF?

Point _____

F
•

• T

I •

SRB
228

3 Which number sentence is NOT true? Choose the best answer.

Ⓐ $5,389 > 3,389$

Ⓑ $70,642 < 70,699$

Ⓒ 1 million $= 1,000,000$

Ⓓ $800,032 = 8$ hundred 32 thousand

Ⓔ $400 + 30 + 5 < 4,000 + 30 + 5$

SRB
81

4 Subtract using U.S. traditional subtraction.

a. $\begin{array}{r} 2,231 \\ -\ 1,084 \\ \hline \end{array}$ b. $\begin{array}{r} 5,603 \\ -\ 3,466 \\ \hline \end{array}$

SRB
100-101

5 Chloe is replacing 3 different sections of the fence in her backyard. She needs fencing in lengths of 13 yards, 17 yards, and 20 yards. How much fencing does she need in feet?

Answer: _____ feet

SRB
186-187

6 Use a straightedge to draw a right angle.

SRB
229

40

Finding Multiples

1 List the first 5 multiples of 10. _____

SRB
55

2 List the first 10 multiples of 6. _____

3 Name 5 multiples of 7. _____

4 Name 5 multiples of 9. _____

5 **a.** List the first 10 multiples of 4. _____

   b. List the first 10 multiples of 8. _____

   c. Which multiples of 4 are also multiples of 8? _____

   d. What pattern do you see when you look at the multiples of 8 and 4?

   _____

   _____

6 The numbers 15, 25, and 40 are all multiples of ____.

7 The numbers 3, 6, and 15 are all multiples of ____.

8 Is 57 a multiple of 7? _____ Explain how you know. _____

   _____

9 Is 81 a multiple of 9? _____ Explain how you know. _____

   _____

10 **a.** List the factors of 12. List the multiples through 12 of each factor.

Factors of 12	Multiples of the Factors (of 12)

   b. Is 12 a multiple of each of its factors? _____ Explain. _____

   _____

41

Place Value through Hundred-Thousands

Use your knowledge of place value to solve.

SRB
78-81

1 Write these numbers in order from least to greatest.

964 9,460 96,400 400,960 94,600

_____

_____

_____

_____

_____

2 **a.** A number has . . .

5 in the hundreds place

7 in the ten-thousands place

0 in the ones place

2 in the hundred-thousands place

9 in the thousands place

8 in the tens place

Write the number.

____ ____ ____ , ____ ____ ____

b. Write the number in words.

_____

_____

3 Write < or > to make each number sentence true.

a. 941 ____ 491

b. 7,023 ____ 70,023

c. 243,342 ____ 241,339

d. 903,710 ____ 907,031

4 What is the value of the digit 8 in the numbers below?

a. 807,941 _____

b. 583 _____

c. 8,714 _____

d. 86,490 _____

5 Write each number using digits.

a. fifteen thousand, two hundred ninety-seven _____

b. four hundred eighty-seven thousand, sixty-three _____

6 Write 87,320 in expanded form.

_____

Math Boxes

1 Convert the following measures.

a. 18 yards = _____ feet

b. 10 feet = _____ inches

c. 25 yards = _____ feet

d. 5 yards 2 feet = _____ feet

SRB
186-187

2 Ari's backyard contains about an acre of lawn shaped in a rectangle. The rectangle measures 660 feet long by 66 feet wide. If Ari measures the distance around the edge, about how many feet is it?

Number model:

_____

_____

Answer: _____ feet

SRB
92-93,
200-201

3 Fill in the missing digits.

```
    3,   5   9  ☐
+   7,  ☐   0   5
─────────────────
   11,   1   0   1
```

SRB
92-93

4 **a.** Draw a pair of parallel line segments.

b. Draw a pair of perpendicular line segments.

SRB
230-231

5 **Writing/Reasoning** Justin's incorrect answer to Problem 2 was 726 feet. Explain what he might have done to get that answer.

_____

_____

_____

SRB
92-93,
200-201

Math Boxes

43

Classifying Numbers as Prime or Composite

SRB
53-54

A **prime number** has exactly two different factors: 1 and the number itself.

A **composite number** has more than two different factors.

1 List all the factors for each number in the table. Write *P* if the number is a prime number or *C* if the number is a composite number.

Number	Factors	P or C	Number	Factors	P or C
2	1, 2	P	21		
3			22		
4	1, 2, 4	C	23		
5	1, 5	P	24		
6			25		
7			26		
8			27		
9			28		
10	1, 2, 5, 10	C	29		
11	1, 11	P	30		
12			31		
13			32		
14			33		
15			34		
16	1, 2, 4, 8, 16	C	35		
17			36		
18			37		
19			38		
20			39		

2 How many factors does each prime number have? _____

3 Can a composite number have exactly 2 factors? _____

44

Factor Captor Strategies

Work alone to answer the questions below. Then compare your answers to your partner's. If your answers don't agree with your partner's answers, try to convince your partner that your answers are correct.

1	2	3	4	5	6	7	8	9	10
11	12	13	14	15	16	17	18	19	20
21	22	23	24	25	26	27	28	29	30

① Suppose you played *Factor Captor* using the number grid above. No numbers have been covered yet. Which is the best number choice you could make? Why?

_____

_____

_____

② Suppose the 29 and 1 squares have already been covered. Which is the best number choice you could make? Why?

_____

_____

_____

③ Suppose that the 29, 23, and 1 squares have already been covered. Which is the best number choice you could make? Why?

_____

_____

_____

Math Boxes

1 Complete.

   a. Name all the factors of 12.

   _____, _____, _____,

   _____, _____, _____

   b. Name the factor pairs of 15.

   _____ and _____

   _____ and _____

   SRB
   53

2 To find the area of a rectangle, use the formula:

   $A = l * w$

   What is the area of a rectangle with a width of 6 inches and a length of 12 inches?

   Equation: _____

   Area: _____ square inches

   SRB
   204

3 Round 849,897 to the nearest thousand. Fill in the circle next to the correct answer.

   Ⓐ 840,000

   Ⓑ 900,000

   Ⓒ 850,000

   Ⓓ 800,000

   SRB
   85-87

4 **a.** Name 3 multiples of 7.

   _____

   b. Name 3 multiples of 6.

   _____

   c. Name 3 multiples of 8.

   _____

   SRB
   55

5 Circle the parallel lines.

   SRB
   230

6 Mimi lives down the street from her friends Lisa and Ellen. It is 24 yards from Mimi's house to Lisa's house and 37 more yards from Lisa's house to Ellen's house. How far is it from Mimi's house to Ellen's house in feet?

   Answer: _____ feet

   SRB
   186-187

46

"What's My Rule?"

Mr. Cheng's class is trying to figure out the rule for the table below. For the rule to be correct, the rule must work for all the rows.

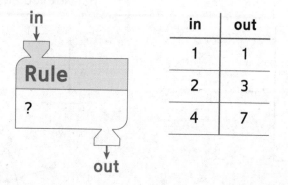

in	out
1	1
2	3
4	7

In the table below, the first column shows educated guesses, or **conjectures,** for rules that Mr. Cheng's students made. Some rules are correct and some are not.

Circle *Yes* or *No* to tell whether the rule is correct. Then write an explanation, or **argument,** for why you think the rule is correct or not.

Conjecture for Rule	Correct? (Circle *Yes* or *No*)	Argument
Multiply by 1.	Yes No	
Add 3.	Yes No	
Double the number you put in and subtract 1.	Yes No	

47

Math Boxes

Math Boxes

1 Find the area and perimeter of the rectangle.

7 in.

4 in.

Area = _____ square inches

Perimeter = _____ inches

SRB
200,
204

2 Circle the sets of lines that appear to be perpendicular to each other.

SRB
230

3 Dan has two dogs, Jango and Riley, and a cat, Kisa. Every year, Jango eats 1,793 dog treats, Riley eats 2,094 treats, and Kisa eats 1,014 cat treats.

Estimate how many more treats both dogs eat than the cat.

Estimate:

_____

SRB
82-84

4 **a.** Name 8 prime numbers under 20.

_____

b. Name 10 composite numbers under 20.

_____

_____

SRB
54

5 **Writing/Reasoning** How do you know the numbers in Problem 4a are prime?

SRB
54

48

Units of Time

Use the measurement scales to fill in the tables.

SRB
198-199

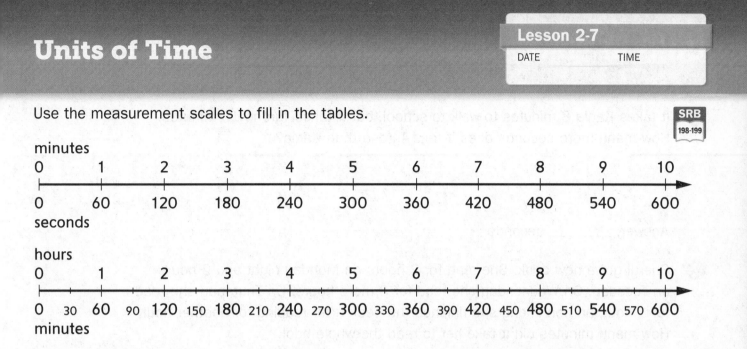

minutes

| 0 | 1 | 2 | 3 | 4 | 5 | 6 | 7 | 8 | 9 | 10 |

| 0 | 60 | 120 | 180 | 240 | 300 | 360 | 420 | 480 | 540 | 600 |

seconds

hours

| 0 | 1 | 2 | 3 | 4 | 5 | 6 | 7 | 8 | 9 | 10 |

0 30 60 90 120 150 180 210 240 270 300 330 360 390 420 450 480 510 540 570 600

minutes

①

Hours	Minutes
1	
2	
3	
5	
7	

②

Minutes	Seconds
1	
4	
5	
8	
9	

③

Hours	Minutes
	600
15	
21	
40	
60	

④

Minutes	Seconds
11	
14	
22	
	1,500
30	

Solve the problems. Use the scale at the top of the page or draw your own.

⑤ Hal needed to do some errands. It took him 30 minutes to grocery shop,
15 minutes to check out books at the library, and 10 minutes to pick up
the dry cleaning. How many seconds did he spend doing errands?

Answer: _____ seconds

49

6 It takes Kania 8 minutes to walk to school, but it takes Tyrone 18 minutes. How many more seconds does it take Tyrone than Kania?

Answer: _____ seconds

7 Sheryl got a new book. She read for 2 hours on Monday night and 2 hours on Tuesday. On Wednesday she read for 3 more hours. On Thursday she read for 1 hour, and on Friday she finished the book after reading for another hour. How many minutes did it take her to read the whole book?

Answer: _____ minutes

8 The Williams family decided to visit some state parks. They drove for 3 hours to get to the first one. The family drove for 4 hours the next day, arriving at the next state park. The following day they drove for 5 hours, and on the last day they drove for 4 hours. How many minutes did they drive all together?

Answer: _____ minutes

9 Jayda's mom said that dinner would be ready in 5 minutes. Demond's dad said dinner would be ready in 8 minutes. How many seconds longer does Demond have to wait for dinner than Jayda?

Answer: _____ seconds

10 It took Aaron 3 minutes and 15 seconds to pack his backpack every day this week (5 days). How many seconds is that in all?

Answer: _____ seconds

Lines, Line Segments, and Rays Practice

Use a straightedge to complete the problems.

SRB
228,
230-231

 Draw line segment *AB*.

2 Draw line *GH*.

3 Label point *T* on line *AB*.

 Draw ray *EF*.

5 Draw angle *XYZ* and label each point.

6 Draw 2 sets of parallel lines and label 2 points on each line.

 Draw 2 sets of perpendicular lines and label 2 points on each line.

Math Boxes

1 Complete.

a. Name all the factors of 40.

_____, _____, _____

_____, _____, _____

_____, _____

b. Name all the factor pairs of 36.

_____ and _____

_____ and _____

_____ and _____

_____ and _____

_____ and _____

SRB
53

2 Find the area using $A = l * w$.

Mr. Janacek's class is doing an art project with different-colored squares. How many 1-inch squares can be cut from an 18-inch by 24-inch piece of construction paper?

Answer: _____ squares

SRB
204

3 Round these numbers to the nearest thousand.

a. 3,496 _____

b. 52,743 _____

c. 697,654 _____

d. 999,502 _____

SRB
85-87

4 Find four multiples of 9. Fill in the circle next to all that apply.

Ⓐ 18, 28, 35, 44

Ⓑ 18, 27, 36, 47

Ⓒ 18, 24, 32, 45

Ⓓ 18, 27, 36, 45

Ⓔ 18, 36, 54, 81

SRB
55

5 Draw two sets of parallel lines. Use a straightedge.

SRB
230

6 Arlo's room is 14 feet long. His sister's room is 12 feet long. What is the total length of both rooms in inches?

Answer: _____ inches

SRB
186-187

52

Multiplicative Comparisons

Complete the problems below.

1 42 is 6 times as much as 7. Record as an equation.

_____

2 What number is 4 times as much as 9? Record as an equation. _____

Solve each number story. Write an equation and use a letter to stand for the unknown.

3 Jon's marble rolled 48 inches. Alana's marble rolled 12 inches. Jon's marble rolled how many times as far as Alana's?

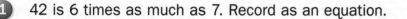

48 inches

12 inches

a. Write an equation with an unknown to represent this comparison.

_____

b. Answer: _____ times as far

4 Samara has 5 times as many apples as Asher. If Asher has 9 apples, how many does Samara have?

a. Equation with unknown: _____

b. Answer: _____ apples

5 Sally is 21 years old. Tonya is 3 times as old as Sally. How old is Tonya?

a. Equation with unknown: _____

b. Answer: _____ years old

6 Write a comparison number story using the equation $8 * 5 = 40$.

_____

_____

_____

Math Boxes

1 Find the area and perimeter of the rectangle below.

60 ft

80 ft

Area = _____ square feet

Perimeter = _____ feet

SRB 200, 204

2 **a.** Draw a set of perpendicular lines.

b. Draw intersecting lines that are not perpendicular.

SRB 230

3 Ted runs a landscape business and needs at least 850 tulip bulbs for fall planting. The plant nursery sent several packages of bulbs: 3 packs with 33 bulbs each, 5 packs with 18 bulbs each, 6 packs with 52 bulbs each, and 3 packs with 105 bulbs each. Estimate how many bulbs Ted received.

Estimate: _____

SRB 82-84

4 Write *T* for true or *F* for false.

_____ **a.** Every composite number has at least 3 factors.

_____ **b.** A composite number is always an even number.

_____ **c.** A prime number can be a composite number.

_____ **d.** 1, 4, 8, and 9 are all composite numbers.

SRB 54

5 **Writing/Reasoning** Renee said that the landscaper in Problem 3 has enough bulbs.

Do you agree? Explain. _____

_____

_____

SRB 82-84

54

Multiplicative Comparison Problems

Use a diagram or drawing to show the relationship between quantities, if needed. Write an equation with an unknown to represent and solve each number story.

SRB
56-57

① It takes 3 hours to fly from Chicago to Alberta, Canada. It takes 4 times that amount to fly from Chicago to Buenos Aires, Argentina. How many hours does it take to fly to Buenos Aires?

Equation with unknown: _____ Answer: _____ hours

② Shar used her chalk to draw a line that was 6 feet long. Diego's line was 36 feet long. Diego's line was how many times as long as Shar's?

Equation with unknown: _____ Answer: _____ times as long

③ Complete the table.

Comparison Statement	Equation	Diagram
a. 4 times as many as 7		
b.	6 * 5 = 30	
c.		$R =$ ⬚ 11 ⬚ | | | $K =$ ⬚ 4 times as many as Randee ⬚ R is the number of T-shirts Randee sold. K is the number of T-shirts Kristin sold.

Math Boxes

1 The number 24 has how many factors? Fill the circle next to the correct answer.

(A) 10 factors

(B) 7 factors

(C) 8 factors

(D) 5 factors

(E) 2 factors

SRB
53

2 Maeve is 3 years old. Maeve's brother is 4 times as old as Maeve. Write an equation with an unknown to represent this problem. Then solve.

Equation with unknown: _____

How old is Maeve's brother?

_____ years old

SRB
37, 47, 56-57

3 Television coverage of the hockey game began at 7 P.M. The game went into three overtimes and did not end until midnight. How many minutes did the game last?

Answer: _____ minutes

SRB
198-199

4 Benji gave away his entire set of 850 plastic blocks. He gave 358 blocks to Carl and 267 blocks to Beth. How many blocks did he have left to give to Nico?

Estimate: _____

Answer: _____ blocks

Number model with answer:

_____

Compare your answer to your estimate. Does your answer make sense?

_____

SRB
47, 83

5 **Writing/Reasoning** For Problem 2, Abby said Maeve's brother is 15. She explained that she added three years to Maeve's age four times, so $3 + 3 + 3 + 3 + 3 = 15$.

Do you agree with her reasoning? _____ Why or why not?

_____

_____

SRB
56-57

Math Boxes

Properties of Triangles

Name an angle property of each triangle. Circle any right angles.
Then name the type of triangle.

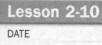

Right triangle: has one right angle
Acute triangle: has angles all smaller than a right angle
Obtuse triangle: has one angle larger than a right angle

1

Property: _____

_____

Type of triangle: _____

2

Property: _____

_____

Type of triangle: _____

3

Property: _____

_____

Type of triangle: _____

4

Property: _____

_____

Type of triangle: _____

Try This

5   Now try your own. Draw a triangle.
Name a property and the type of triangle.

Math Boxes
Preview for Unit 3

Math Boxes

① Use your fraction circles to solve.

How many dark green fraction pieces does it take to cover a red circle?

_____ dark green pieces

SRB
130

② Divide the rectangle into halves. Use a straightedge.

Use dashes to divide the rectangle into halves another way.

SRB
125-126

③ Circle the larger fraction.

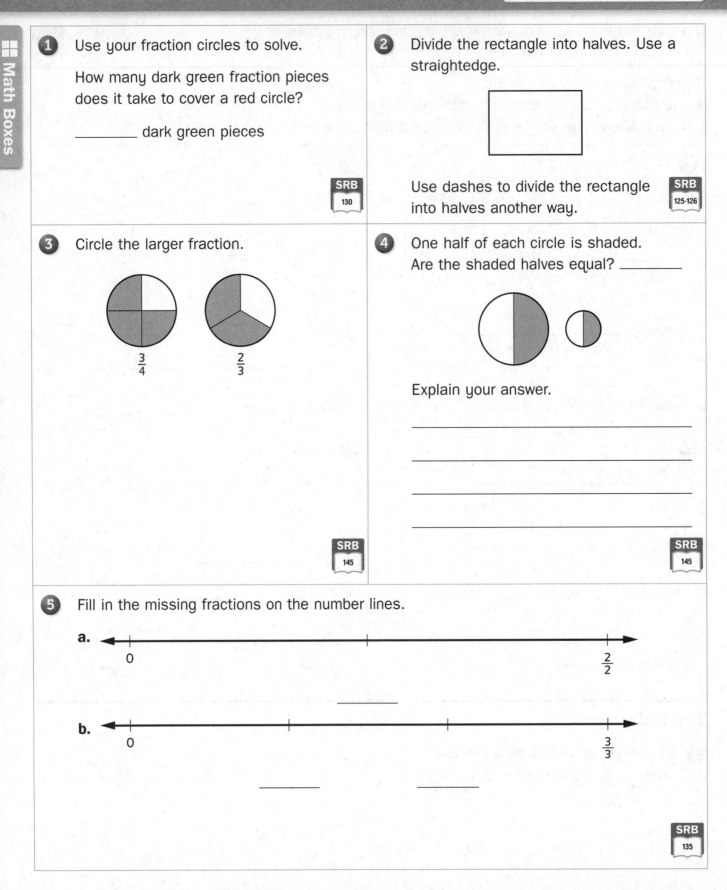

$\frac{3}{4}$ $\frac{2}{3}$

SRB
145

④ One half of each circle is shaded. Are the shaded halves equal? _____

Explain your answer.

_____

_____

_____

_____

SRB
145

⑤ Fill in the missing fractions on the number lines.

a.

0 $\frac{2}{2}$

_____

b.

0 $\frac{3}{3}$

_____   _____

SRB
135

Properties of Quadrilaterals

1 Sort the quadrilaterals from *Math Masters* page TA18 into 3 groups. Then describe the property you used to form each group.

SRB
234-235

a. Write the letters of the quadrilaterals that go in each group.

Group One	Group Two	Group Three
A,	H,	E,

b. All of the quadrilaterals in Group One have this property in common:

_____

c. All of the quadrilaterals in Group Two have this property in common:

_____

d. All of the quadrilaterals in Group Three have this property in common:

_____

2 In which of your groups does quadrilateral X belong? _____ Explain.

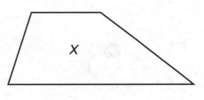

Try This

3 Using your Geometry Template, draw a quadrilateral below and explain in which group it belongs.

Math Boxes

1 Complete the "What's My Rule?" table and state the rule.

Rule: _____

In	Out
5	25
6	36
	49
8	
	81

SRB
65-67

2 What is the value of the digit **4** in each number?

a. 154 _____

b. 349 _____

c. 547,326 _____

d. 84,391 _____

e. 3,472 _____

f. 473,962 _____

SRB
78-79

3 Bianca's dining room measures 10 feet by 18 feet. Her family plans to buy new wall-to-wall carpeting for it. How many square feet of carpet do they need to order?

Area: _____ square feet

SRB
204

4 Tanisha is writing a 600-word story for school. She has written 3 pages; page 1 has 135 words, page 2 has 142 words, and page 3 has 85 words. How many more words does she need to write?

Estimate: _____

Answer: _____ words

Number model with answer:

_____

Compare your answer to your estimate. Does your answer make sense?

_____

SRB
47, 83

5 **Writing/Reasoning** In Problem 2 the digit **4** has many different values. Describe the pattern you see in the answers and explain how the value of the digit 4 changes from problem to problem.

SRB
78-79

_____

_____

_____

Identifying Line Symmetry

1 Cut out the drawings on *Math Masters*, page 84. Fold them to find lines of symmetry. Record your answers to the right.

Object	Number of Lines of Symmetry
leaf	
football	
turtle	
bow tie	

SRB
238

2 Cut out each polygon on *Math Masters*, pages 85 and 86. Fold to find all the lines of symmetry for each polygon. Record the results below.

Polygon	Number of Lines of Symmetry	Polygon	Number of Lines of Symmetry
A		F	
B		G	
C		H	
D		I	
E		J	

3 Study the results in the tables above.

a. How many lines of symmetry are in a pentagon with 5 equal sides (Polygon I)?

_____ lines

b. How many lines of symmetry are in a hexagon with 6 equal sides (Polygon J)?

_____ lines

c. How many lines of symmetry would you expect in an octagon with 8 equal sides?

_____ lines

4 Circle the polygons that are symmetrical.

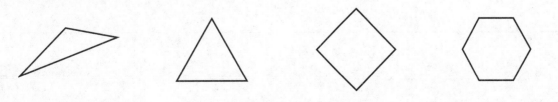

Math Boxes

1 Complete.

  a. How many different factors
  does a prime number have? _____

  b. Give the factor pairs of 48.

  _____

  _____

  _____

  SRB
  53-54

2 Monica has saved 5 times as much
money as her sister. Monica has
saved $100. How much money has
her sister saved?

Equation with unknown:

_____

Answer: $_____

  SRB
  37, 47

3 Convert between units of time to answer
the questions below.

  a. It took Ely 4 hours to put a 500-piece
  puzzle together. How many minutes
  is that?

  _____ minutes

  b. It snowed for 10 hours. How many
  seconds is that?

  _____ seconds

  SRB
  198-199

4 Sumi and her sister combined their sticker
collections. Sumi had 374 stickers and her
sister had 193. Their cousin Sayuri has
743 stickers. How many more stickers
does Sayuri have than the sisters?

Estimate: _____

Answer: _____ stickers

Number model with answer:

_____

Compare your answer to your estimate.
Does your answer make sense?

_____

  SRB
  47, 83

5 **Writing/Reasoning** How do you know you found all the factor pairs for 48
in Problem 1?

_____

_____

_____

_____

  SRB
  53

"What's My Rule?"

Complete the "What's My Rule?" tables and explain the patterns that you find.

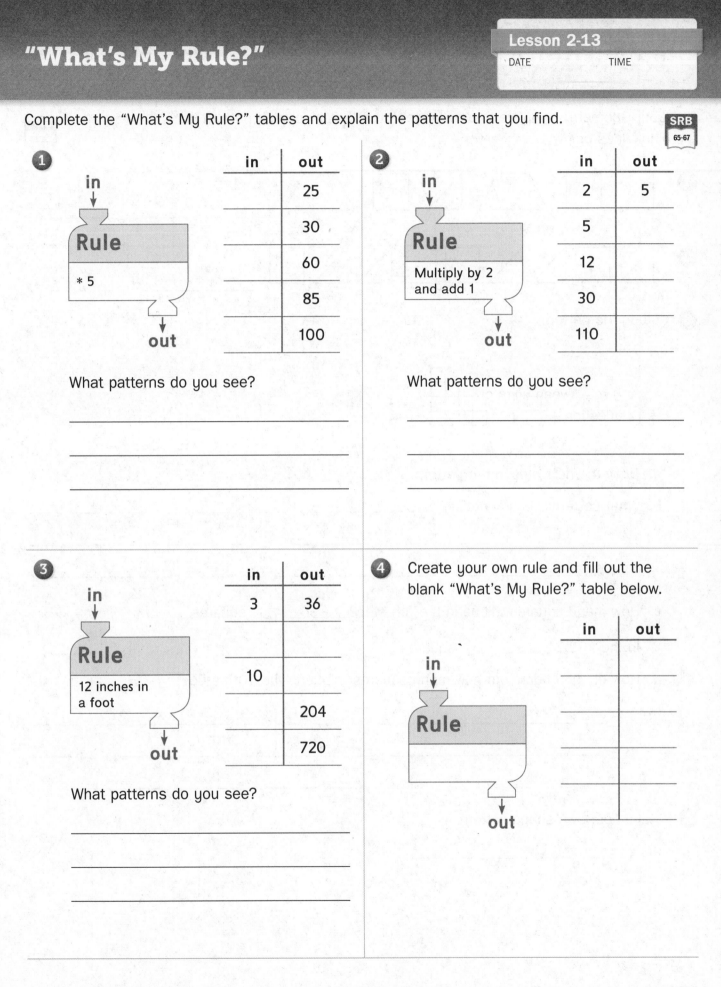

1

in
↓

Rule

* 5

↓
out

in	out
	25
	30
	60
	85
	100

What patterns do you see?

_____

_____

_____

2

in
↓

Rule

Multiply by 2
and add 1

↓
out

in	out
2	5
5	
12	
30	
110	

What patterns do you see?

_____

_____

_____

3

in
↓

Rule

12 inches in
a foot

↓
out

in	out
3	36
10	
	204
	720

What patterns do you see?

_____

_____

4 Create your own rule and fill out the
blank "What's My Rule?" table below.

in
↓

Rule

↓
out

in	out

63

Shape Patterns

Use your Geometry Template to draw shapes that continue each pattern on the blank lines below.

SRB
58-60

1. ◁ ☐ ⬜ ◁ ☐ ⬜ _____ _____ _____

2. ◺ ☐ ◺ ◺ ☐ ◺ ◺ ◺ ☐ _____ _____ _____ _____

3. Study the pattern.

☐   ☐☐   ☐☐   ☐☐☐   _____
         ☐   ☐☐☐ ☐☐☐☐
1   2   3   4   5

   a. Draw the next step in the pattern.

   b. What patterns do you notice? _____

   _____

   _____

   c. How many squares will be in the 6th step? _____ squares

   In the 10th? _____ squares

   d. How did you figure out how many squares will be in the 10th step?

   _____

   _____

   _____

4. Make your own shape pattern.

   Rule: _____

64

Adding and Subtracting

Make an estimate. Write a number model to show your thinking. Then solve using U.S. traditional addition or subtraction. Compare your answer with your estimate to see if your answer makes sense.

SRB
92-93,
100-101

1)
```
   2, 7 4 8
+  8, 6 7 9
```

Estimate: _____

2)
```
   9, 5 9 8
-  5, 3 5 9
```

Estimate: _____

3) 5,392 + 1,175 = _____

Estimate: _____

4)
```
   1, 7 2 8
-  1, 3 8 2
```

Estimate: _____

5)
```
   4, 9 2 3
+  5, 2 3 4
```

Estimate: _____

6) 7,000 − 5,093 = _____

Estimate: _____

Math Boxes

1 Complete the "What's My Rule?" table and state the rule.

Rule: _____

in	out
300	
	150
400	200
450	
500	300

SRB
65-67

2 Write *T* for true or *F* for false.

____ The place value of each digit in a number is ten times as large as the place value to the right.

____ 10 [10s] is equal to 1,000.

____ 10 [100s] is equal to 1,000.

____ The place value of each digit in a number is one hundred times as large as the place value to its right.

SRB
78-79

3 Dasad is painting a wall in his bedroom. The wall measures 9 feet by 14 feet. Which number model shows how many square feet Dasad has to paint?

Choose ALL that apply.

(A) $9 + 9 + 14 + 14 = f$

(B) $f = 21 * 21$

(C) $14 * 9 = f$

(D) $9 * 9 * 14 * 14 = f$

(E) $f = 9 * 14$

SRB
204

4 The teacher used 76 sheets of paper for the math activity, 48 sheets for the reading activity, and 126 sheets for the science activity. How many sheets are left from the 500-page pack?

Estimate: _____

Answer: _____ sheets

Number model with answer:

_____

Compare your answer to your estimate. Does your answer make sense?

SRB
47, 83

5 **Writing/Reasoning** Look at the completed "What's My Rule?" table in Problem 1. Describe any patterns you see in the *out* column.

_____

_____

_____

_____

SRB
58-60

Math Boxes
Preview for Unit 3

1 Use your fraction circles to solve.

How many purple fraction pieces does it take to cover a red circle?

_____

SRB
130

2 Which figure shows $\frac{1}{3}$ shaded? Fill in the circle next to ALL that apply.

Ⓐ   Ⓑ   Ⓒ

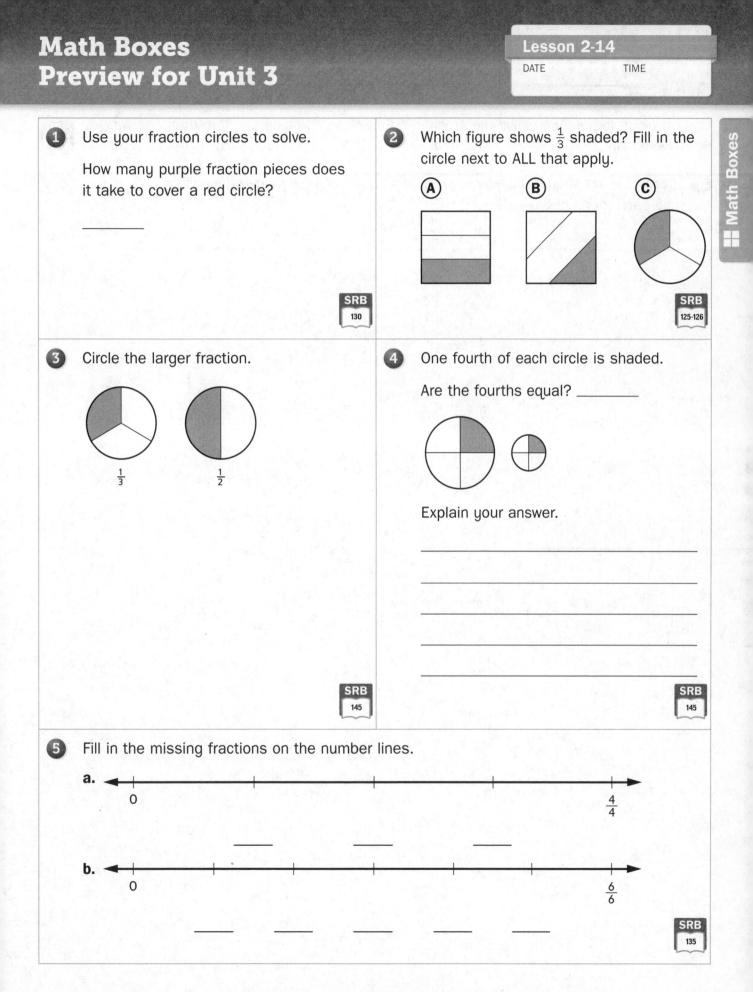

SRB
125-126

3 Circle the larger fraction.

$\frac{1}{3}$ $\frac{1}{2}$

SRB
145

4 One fourth of each circle is shaded.

Are the fourths equal? _____

Explain your answer.

_____

_____

_____

_____

_____

SRB
145

5 Fill in the missing fractions on the number lines.

a. 0 $\frac{4}{4}$

_____   _____   _____

b. 0 $\frac{6}{6}$

_____ _____ _____ _____ _____

SRB
135

Math Boxes

67

Equal Sharing

Use drawings to help you solve the problems. Solve each problem in more than one way. Show your work.

SRB
124-125,
156-157

1. Three friends shared 4 chicken quesadillas equally. How many quesadillas did each friend get?

   _____ quesadillas

   One way:

   Another way:

68

Equal Sharing (continued)

2 A group of 4 soccer players brought 6 gallons of water to the tournament on Saturday. If the players shared the water equally, how many gallons of water did each player get?

_____ gallons of water

One way:

Another way:

Try This

3 The team played 3 games during the tournament. One player decided to drink the same amount of water at each game. How many gallons of water did the player drink at each of the 3 games?

_____ gallon(s)

69

Math Boxes

1 Write each number using digits.

   a. Three thousand, six _____

   b. Ten thousand, fourteen _____

   c. Seven hundred fourteen thousand, three hundred two _____

   d. One million, seventy-six thousand, ninety _____

SRB 78-79

2 Round each number to the nearest 1,000.

   a. 825 _____

   b. 72,804 _____

   c. 321,549 _____

   d. 204,341 _____

   e. 6,000,487 _____

SRB 85-87

3 Which two statements represent the equation 16 * 2 = 32?

  ☐ 32 is 16 times as many as 2.

  ☐ 16 is 2 times as many as 32.

  ☐ 32 is 16 more than 2.

  ☐ 32 is 2 times as many as 16.

SRB 56-57

4 Fill in the blanks.

   a. 5 feet = _____ inches

   b. 8 feet = _____ inches

   c. 3 feet 3 inches = _____ inches

   d. 12 yards = _____ feet

   e. 6 yards 2 feet = _____ feet

   f. 9 hours = _____ minutes

   g. 4 minutes = _____ seconds

SRB 186-187, 198

5 Solve using U.S. traditional addition or subtraction.

   a.
$$\begin{array}{r} 5,967 \\ +\ 7,628 \\ \hline \end{array}$$

   b.
$$\begin{array}{r} 6,545 \\ -\ 4,659 \\ \hline \end{array}$$

SRB 92-93, 100-101

6 Which groups below contain all prime numbers? Fill in the circle next to ALL that apply.

  ○ **A.** 4, 56

  ○ **B.** 13, 89

  ○ **C.** 37, 53

  ○ **D.** 42, 61

SRB 54

Equivalent Names for Fractions

SRB
136-137

Fraction	Equivalent Fractions
$\frac{0}{2}$	~~~~~~~~~~~~~~~~~~~~~~~~~~~~~~~~
$\frac{1}{2}$	$\frac{2}{4}$, $\frac{3}{6}$, $\frac{4}{8}$, $\frac{5}{10}$, $\frac{100}{200}$, $\frac{50}{100}$, $\frac{30}{60}$, $\frac{6}{12}$, $\frac{8}{16}$
$\frac{2}{2}$	~~~~~~~~~~~~~~~~~~~~~~~~~~~~~~~~
✗ $\frac{1}{3}$	$\frac{2}{6}$, $\frac{4}{12}$
✗ $\frac{2}{3}$	$\frac{4}{6}$, $\frac{8}{12}$
✗ $\frac{1}{4}$	$\frac{2}{8}$, $\frac{3}{12}$
$\frac{3}{4}$ ×2 ×?	$\frac{6}{8}$, $\frac{18}{24}$, $\frac{24}{32}$, $\frac{12}{16}$, $\frac{9}{12}$
✗ $\frac{1}{5}$	$\frac{2}{10}$,
$\frac{2}{5}$ ××	$\frac{4}{10}$, $\frac{6}{15}$, $\frac{12}{30}$, $\frac{8}{20}$, $\frac{10}{25}$
$\frac{3}{5}$	$\frac{6}{10}$, $\frac{12}{20}$, $\frac{15}{25}$, $\frac{36}{60}$, $\frac{18}{30}$
$\frac{4}{5}$	
$\frac{1}{6}$	
$\frac{5}{6}$	

71

Math Boxes

---

1 Naomi has 8 dollars. Her sister has 9 times that amount. How much money does her sister have?

Number model with unknown:

_____

Answer: $_____

SRB
56-57

2 I am a multiple of 3, 6, and 9. What number am I? Fill in the circle next to the best answer.

○ **A.** 45

○ **B.** 63

○ **C.** 52

○ **D.** 36

SRB
55

---

3 Complete the table.

Minutes	Seconds
10	
14	
21	
30	
	2,400

SRB
198

4 Write *T* for True or *F* for False.

a. Right triangles can have

2 right angles. ____

b. Right triangles can have

2 acute angles. ____

c. Right triangles can have

2 obtuse angles. ____

d. Right triangles have a 90° angle. ____

SRB
233

---

5 **Writing/Reasoning** When Sharita solved Problem 1, she wrote the following number model: $8 + 9 = d$. Explain why you agree or disagree with the way she solved the problem.

_____

_____

_____

_____

SRB
56-57

72

Fractions on Number Lines

Fill in the missing fractions on the number lines.

SRB
133-135

1 0 ⟵——┼————————┼————————┼————————┼————————┼——⟶
       0              $\frac{1}{4}$ 1

2 0 ⟵——┼————————┼————————┼————————┼————————┼——⟶
       0   1
              ___   ___   ___

3 ⟵——┼————————┼————————┼————————┼————————┼————————┼——⟶
       0 $\frac{4}{6}$ 1
           ___   ___   ___   ___

4 ⟵——┼————————┼————————┼————————┼————————┼————————┼——⟶
       0   1
         ___   ___   ___   ___   ___

5 ⟵——┼————————┼————————┼————————┼————————┼————————┼——⟶
       0   3
           ___ $\frac{2}{2}$ $\frac{3}{2}$ ___   ___

6 ⟵——┼————┼————┼————┼————┼————┼————┼————┼————┼——⟶
       0                              $\frac{5}{5}$ $\frac{8}{5}$ 2
         ___   ___   ___   ___   ___   ___   ___

7 The number lines in Problems 3 and 5 are divided into the same number of intervals.
       Why did you use different fractions to name these intervals?

_____

_____

_____

_____

Equivalent Fractions Poster

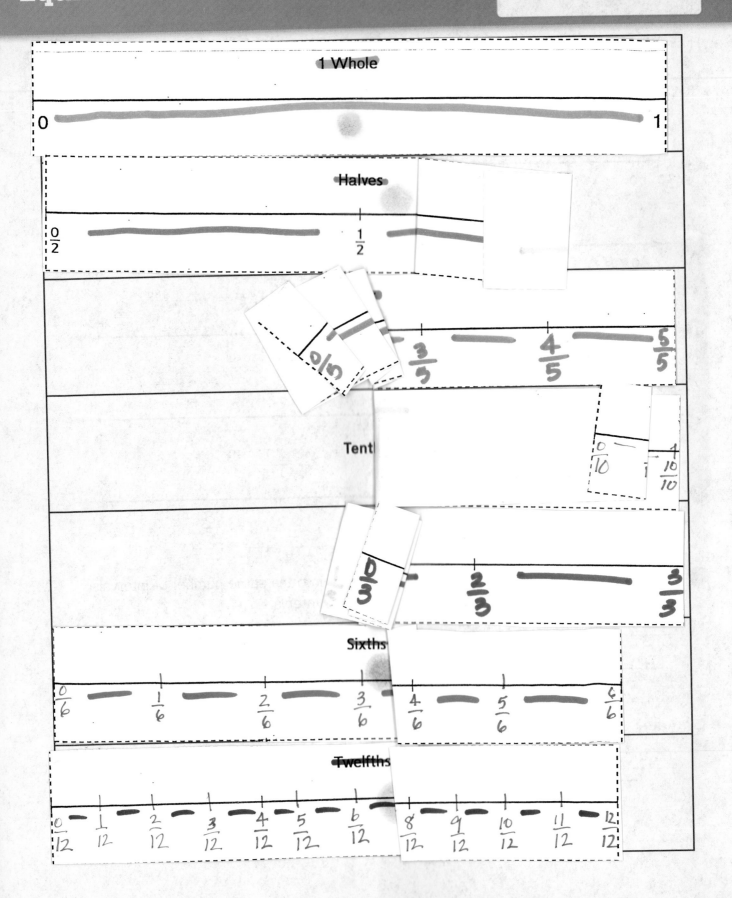

Equivalent Fractions

Use the Equivalent Fractions Poster on journal page 74 to help you
solve the problems below.

SRB
136-137

1 Circle the true number sentences.

$\dfrac{1 \times 2}{5 \times 2} = \dfrac{2}{10}$ (circled)

$\dfrac{3}{6} = \dfrac{4}{12}$

$\dfrac{4}{10} = \dfrac{2}{5}$ (circled)

$\dfrac{1}{2} = \dfrac{5}{10}$ (circled)

$\dfrac{4}{5} = \dfrac{9}{10}$

$\dfrac{1}{5} = \dfrac{1}{3}$

2 Insert = or ≠ to make a true number sentence.

a. $\dfrac{8}{10} \underline{\ =\ } \dfrac{4}{5}$ (circled)

b. $\dfrac{2}{3} \underline{\ =\ } \dfrac{8}{12}$

c. $\dfrac{1}{6} \underline{\ \neq\ } \dfrac{1}{10}$

d. $\dfrac{5}{10} \underline{\ =\ } \dfrac{3}{6}$

3 Fill in the missing number.

a. $\dfrac{1 \times 4}{3 \times 4} = \dfrac{4}{12}$

b. $\dfrac{6}{12} = \dfrac{1}{2}$

c. $\dfrac{3}{5} = \dfrac{6}{10}$

d. $\dfrac{1}{2} = \dfrac{3}{6}$

e. $\dfrac{2}{6} = \dfrac{1}{3}$

f. $\dfrac{8 \times 2}{12 \times 2} = \dfrac{4}{6}$

4 What do you notice about the number and size of the equal parts for the
thirds number line compared to the sixths number line?

_____

_____

_____

Math Boxes

1 Write each number using words.

   a. 670 _____

   _____

   b. 3,590 _____

   _____

   c. 103,004 _____

   _____

SRB
78-79

2 What place is each number rounded to?

   a. 5,689 to 5,690:

   Nearest _____*10*_____

   b. 7,623 to 8,000:

   Nearest _____

   c. 19,487 to 20,000:

   Nearest _____

   d. 25,582 to 25,600:

   Nearest _____

SRB
85-87

3 Write a multiplicative comparison number story for the following number sentence.

$6 * 9 = 54$

_____

_____

_____

SRB
56-57

4 Fill in the blanks.

   a. 9 yd = _____ ft

   b. 10 yd 1 ft = _____ ft

   c. 4 feet 7 inches = _____ in.

   d. 6 hr = _____ min

   e. 8 hr = _____ min

   f. 2 hr 30 min = _____ min

SRB
186-187,
198

5 Solve using U.S traditional addition or subtraction.

   a.   1, 6 5 9   **b.**   7, 9 5 1
       +   4 9 9   − 3, 5 6 1
       _____   _____

SRB
92-93,
100-101

6 Write *prime* or *composite* in the blanks.

   a. A _____ number has exactly 2 different factors.

   b. Both 51 and 63 are _____ numbers.

   c. Both 23 and 67 are _____ numbers.

   d. A _____ number has more than 2 factors.

SRB
54

Equivalent Names for Fractions

Color the circles and write the missing numerators.

SRB
136

Whole
circle

1. Color $\frac{1}{2}$ of each circle.

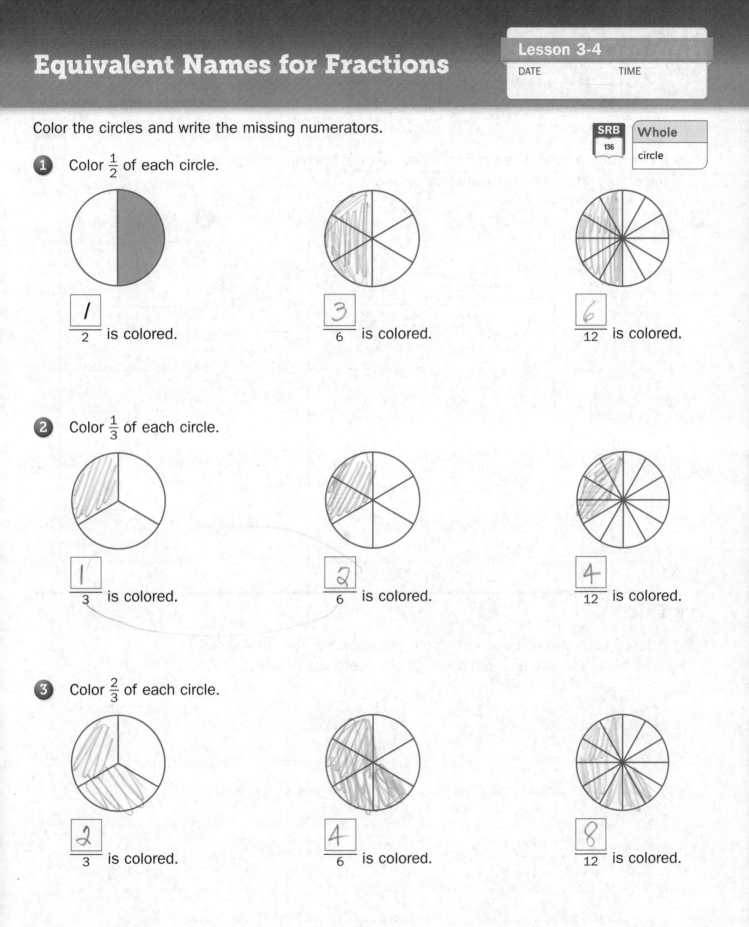

$\dfrac{1}{2}$ is colored.

$\dfrac{3}{6}$ is colored.

$\dfrac{6}{12}$ is colored.

2. Color $\frac{1}{3}$ of each circle.

$\dfrac{1}{3}$ is colored.

$\dfrac{2}{6}$ is colored.

$\dfrac{4}{12}$ is colored.

3. Color $\frac{2}{3}$ of each circle.

$\dfrac{2}{3}$ is colored.

$\dfrac{4}{6}$ is colored.

$\dfrac{8}{12}$ is colored.

Name-Collection Boxes
for Equivalent Fractions

In each name-collection box:

Write the missing number in each fraction so that the fraction belongs in the box.
Fill in the empty boxes with fractions that belong.

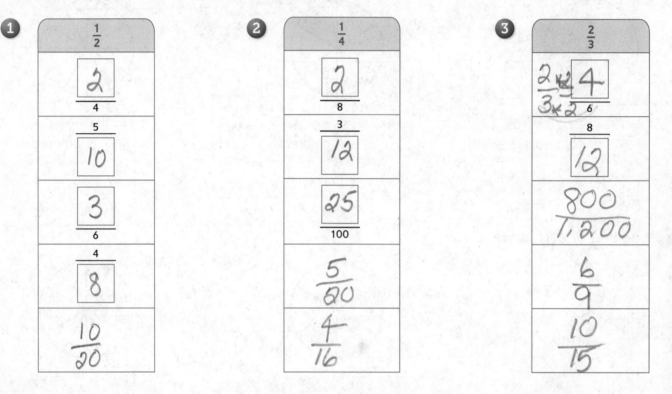

1 $\frac{1}{2}$

$\frac{2}{4}$

$\frac{5}{10}$

$\frac{3}{6}$

$\frac{4}{8}$

$\frac{10}{20}$

2 $\frac{1}{4}$

$\frac{2}{8}$

$\frac{3}{12}$

$\frac{25}{100}$

$\frac{5}{20}$

$\frac{4}{16}$

3 $\frac{2}{3}$

$\frac{2 \times 2}{3 \times 2} = \frac{4}{6}$

$\frac{8}{12}$

$\frac{800}{1,200}$

$\frac{6}{9}$

$\frac{10}{15}$

Try This

4 Make up your own name-collection box problems like the ones above.
Ask a friend to solve your problems. Check your friend's work.

a.

b.

Solving Multistep Number Stories

Estimate and write a number model to show your work. Solve each multistep number story. **SRB** *47, 82-89*

1 Each day a company delivers newspapers to the town of Wayland. It has 158 customers on the north side of town, 378 customers on the west side, and 237 customers on the south side. The company receives 900 newspapers to deliver. How many will be left over?

   a. Estimate: $900 - (200 + 400 + 200) = 100$

   b. Answer: _127_ newspapers

   c. Number model with unknown: $900 - (158 + 378 + 237) = n$

   d. Explain how you know your answer is reasonable.

   My estimate is close to my answer because I rounded to the nearest 100.

2 The following are populations from the 2010 census of five towns on Long Island, New York:

   New Hyde Park Village: 9,712; Old Bethpage: 5,523; Oyster Bay: 6,707; Wading River: 7,719; Riverhead Town: 33,506. How many more people live in Riverhead Town than in the other four towns combined?

   a. Estimate: $34,000 - (10,000 + 6,000 + 7,000 + 8,000) = 3,000$

   b. Answer: _3,845_ people   31,000

   c. Number model with unknown: $33,506 - (9,712 + 5,523 + 6,707 + 7,719) = p$

3 Edward bought 3 packages of hot dogs with 8 hot dogs in each package. He also bought 3 bags of hot dog buns with 10 buns in each bag. Six people were having lunch at his home, including Edward. How many hot dogs with buns could each of them have?

   a. Estimate: $20 ÷ 5 = 4$ 24   30

   b. Answer: _4_ hot dogs with buns

   c. Number model(s) with unknown: $24 ÷ 6 = h$

   d. How many buns would Edward have left over? Answer: _6_ buns

Math Boxes

1 A rectangular fish tank holds 12 fish. A round fish bowl holds 3 fish. How many times as many fish does the rectangular tank hold as the round bowl?

Equation with unknown:

_____

Answer: _____ times as many fish

SRB
56-57

2 Use what you know about multiples to solve the riddles.

a. I am the smallest multiple of 5, 6, and 10. What number am I?

_____

b. I am the smallest multiple of 4, 8, and 16. What number am I?

_____

SRB
55

3 Complete the table.

Hour(s)	Minutes
1	
5	
6	
	480
2 hours 5 minutes	

SRB
198

4 Draw 2 different right triangles. Circle the right angles.

SRB
233

5 **Writing/Reasoning** How do you know the triangles you drew for Problem 4 are right triangles even though they are not the same?

_____

_____

_____

_____

_____

SRB
233

Using Benchmarks

Each of the rectangles below is one whole unit. Label each strip with one of the following fractions: $\frac{1}{10}$, $\frac{9}{10}$, $\frac{7}{8}$, $\frac{4}{7}$, $\frac{2}{5}$, or $\frac{1}{8}$. Use benchmarks of 0, $\frac{1}{2}$, and 1 to help you.

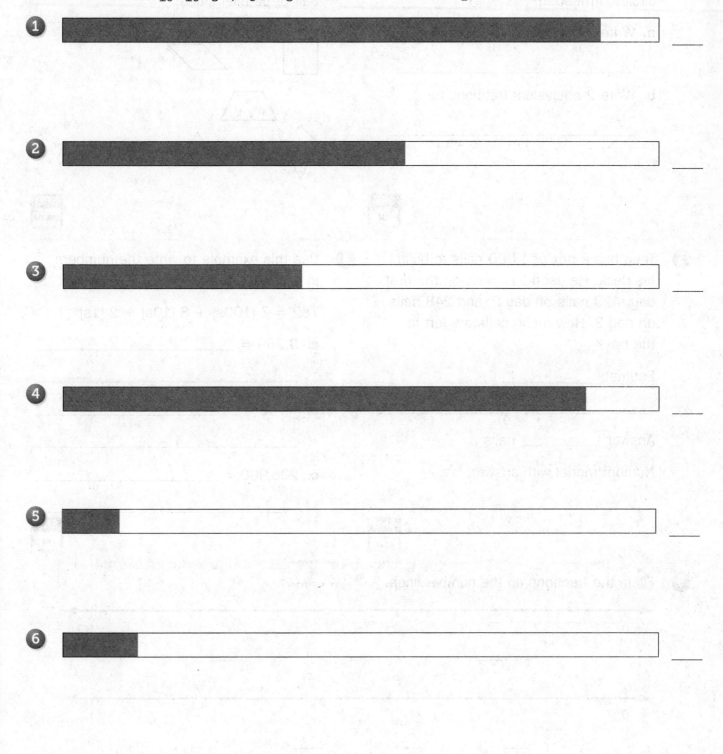

Math Boxes

1 Identify equivalent fractions. Use fraction circles, if needed.

   a. Write 2 equivalent fractions for $\frac{1}{3}$.

   _____

   b. Write 2 equivalent fractions for $\frac{1}{4}$.

   _____

SRB
137

2 Circle the shapes that have at least one pair of parallel sides.

SRB
230, 235

3 Taye has a box of 1,000 nails to build his deck. He used 217 nails on the first day, 423 nails on day 2, and 248 nails on day 3. How many nails are left in the box?

Estimate:

_____

Answer: _____ nails

Number model with answer:

_____

SRB
47,
82-89

4 Use this example to write the numbers in expanded form.

782 = 7 [100s] + 8 [10s] + 2 [1s]

   a. 3,269 = _____

   _____

   b. 9,742 = _____

   _____

   c. 206,900 = _____

   _____

SRB
80

5 Fill in the fractions on the number lines.

SRB
133

Comparing Fractions

Quinn, Nina, Diego, Paula, and Kiana were given 4 same-size burritos to share.

Quinn and Nina shared one burrito. Quinn ate $\frac{1}{4}$ of the burrito, and Nina ate $\frac{2}{4}$.

Diego, Paula, and Kiana each ate part of the other burritos. Diego ate $\frac{2}{3}$ of one burrito,
Paula ate $\frac{2}{5}$ of another burrito, and Kiana ate $\frac{5}{6}$ of the third burrito.

1 Who ate more, Diego or Paula? Or did they eat the same amount? _Diego_

How do you know? _They had the same number of_
pieces but Diego's burrito was only cut
into 3 pieces so his 2 were larger.

Write a number model using the symbols >, =, or < to record the comparison.

$\frac{2}{3} > \frac{2}{5}$

2 Who ate more, Diego or Kiana? Or did they eat the same amount? _Kiana_

How do you know? _$\frac{2}{3} = \frac{4}{6}$ which is what Diego ate._
And Kiana ate $\frac{5}{6}$. $\frac{5}{6}$ is more than
$\frac{4}{6}$ or $\frac{2}{3}$.

Write a number model using the symbols >, =, or < to record the comparison.

$\frac{2}{3} < \frac{5}{6}$

3 Who ate more, Diego or Quinn? Or did they eat the same amount? _Diego_

How do you know? _$\frac{1}{4}$ is less than $\frac{1}{2}$. $\frac{2}{3}$ is more_
than $\frac{1}{2}$. So, Diego ate more.

Write a number model using the symbols >, =, or < to record the comparison.

$\frac{1}{4} < \frac{2}{3}$

Least ————————→ greatest

$\frac{2}{3}$, $\frac{1}{4}$, $\frac{2}{4}$, $\frac{2}{5}$, $\frac{5}{6}$ $\frac{1}{4}$, $\frac{2}{5}$, $\frac{2}{4}$, $\frac{2}{3}$, $\frac{5}{6}$

$\frac{1}{2}$

83

Practicing Place Value

Fill in the blank.

1. The 9 in 9,000 is _____ times as large as the 9 in 900.

2. The 5 in 500,000 is _____ times as large as the 5 in 50,000.

3. The 3 in 300,000 is _____ times as large as the 3 in 3,000.

4. The 2 in 20,000 is _____ times as large as the 2 in 200.

Fill in the blank.

5. The value of the digit 2 in 416,321 is _____ times as large as the value of the 2 in 65,382.

6. The value of the digit 9 in 73,980 is _____ times as large as the value of the 9 in 38,459.

7. The value of the digit 9 in 9,423,631 is _____ times as large as the value of the 9 in 5,972,803.

8. The value of the digit 7 in 578,485 is _____ times as large as the value of the 7 in 295,725.

9. The value of the digit 3 in 3,786,012 is _____ times as large as the value of the 3 in 8,537,600.

10. a. Write a 5-digit number in which the digit 4 is worth 4,000.

   _____

   b. Write a number in which the digit 4 is 10 times as large as the 4 you wrote in your answer to 10a.

   _____

11. a. Write a number in which the digit 6 is worth 6 [10,000s].

   _____

   b. Write a number in which the 6 is 100 times as large as the number you wrote in your answer for Problem 11a.

   _____

Math Boxes

Math Boxes

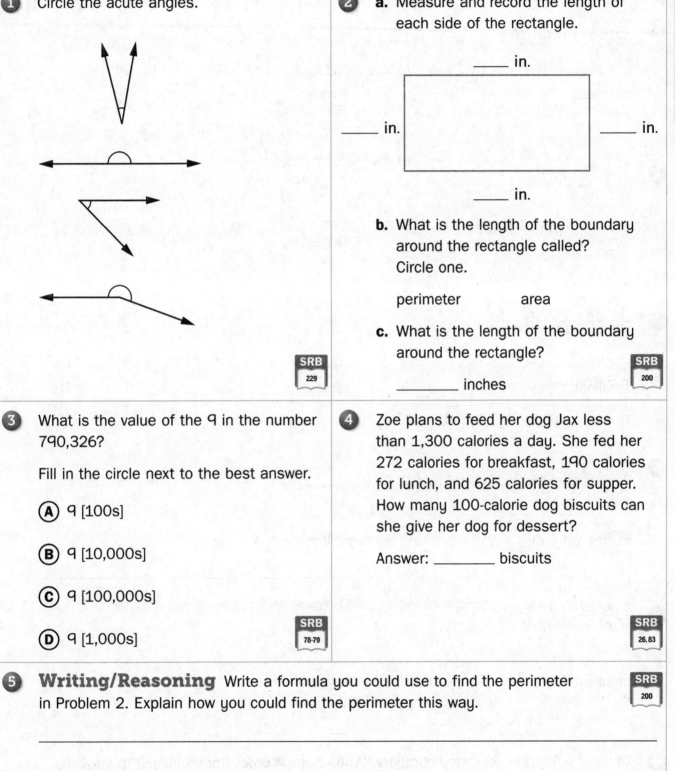

1 Circle the acute angles.

SRB
229

2 **a.** Measure and record the length of each side of the rectangle.

_____ in.

_____ in.   _____ in.

_____ in.

b. What is the length of the boundary around the rectangle called? Circle one.

perimeter   area

c. What is the length of the boundary around the rectangle?

_____ inches

SRB
200

3 What is the value of the 9 in the number 790,326?

Fill in the circle next to the best answer.

(A) 9 [100s]

(B) 9 [10,000s]

(C) 9 [100,000s]

(D) 9 [1,000s]

SRB
78-79

4 Zoe plans to feed her dog Jax less than 1,300 calories a day. She fed her 272 calories for breakfast, 190 calories for lunch, and 625 calories for supper. How many 100-calorie dog biscuits can she give her dog for dessert?

Answer: _____ biscuits

SRB
26, 83

5 **Writing/Reasoning** Write a formula you could use to find the perimeter in Problem 2. Explain how you could find the perimeter this way.

SRB
200

_____

_____

_____

85

Ordering Fractions

Write the following fractions in order from smallest to largest.

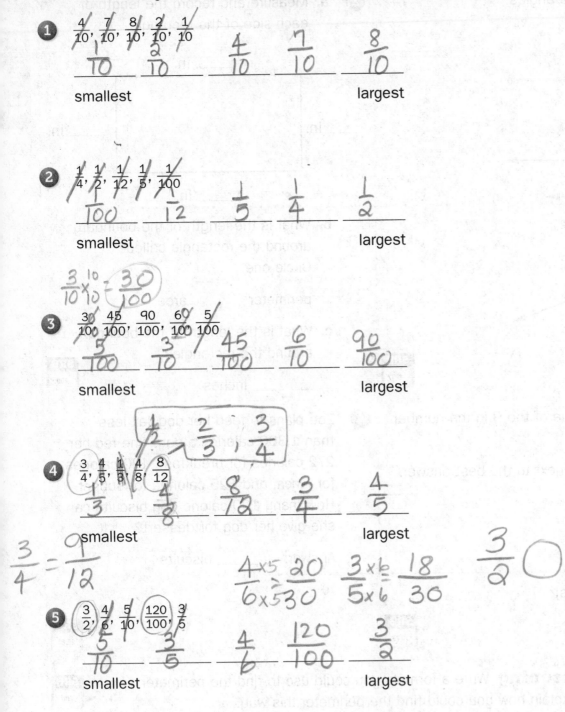

1 $\frac{4}{10}, \frac{7}{10}, \frac{8}{10}, \frac{2}{10}, \frac{1}{10}$

$\frac{1}{10}$ _____ $\frac{2}{10}$ _____ $\frac{4}{10}$ _____ $\frac{7}{10}$ _____ $\frac{8}{10}$ _____

smallest   largest

2 $\frac{1}{4}, \frac{1}{2}, \frac{1}{12}, \frac{1}{5}, \frac{1}{100}$

$\frac{1}{100}$ _____ $\frac{1}{12}$ _____ $\frac{1}{5}$ _____ $\frac{1}{4}$ _____ $\frac{1}{2}$ _____

smallest   largest

$\frac{3}{10} \times \frac{10}{10} = \frac{30}{100}$

3 $\frac{30}{100}, \frac{45}{100}, \frac{90}{100}, \frac{60}{100}, \frac{5}{100}$

$\frac{5}{100}$ _____ $\frac{3}{10}$ _____ $\frac{45}{100}$ _____ $\frac{6}{10}$ _____ $\frac{90}{100}$ _____

smallest   largest

$\frac{1}{2} \quad \frac{2}{3}, \quad \frac{3}{4}$

4 $\frac{3}{4}, \frac{4}{5}, \frac{1}{3}, \frac{4}{8}, \frac{8}{12}$

$\frac{1}{3}$ _____ $\frac{4}{8}$ _____ $\frac{8}{12}$ _____ $\frac{3}{4}$ _____ $\frac{4}{5}$ _____

smallest   largest

$\frac{3}{4} = \frac{9}{12}$

$\frac{4 \times 5}{6 \times 5} = \frac{20}{30} \quad \frac{3 \times 6}{5 \times 6} = \frac{18}{30}$

$\frac{3}{2} \bigcirc \frac{6}{5}$

5 $\frac{3}{2}, \frac{4}{6}, \frac{5}{10}, \frac{120}{100}, \frac{3}{5}$

$\frac{5}{10}$ _____ $\frac{3}{5}$ _____ $\frac{4}{6}$ _____ $\frac{120}{100}$ _____ $\frac{3}{2}$ _____

smallest   largest

6 Choose 5 fractions or mixed numbers. Write them in order from smallest to largest.

_____ _____ _____ _____ _____

smallest   largest

Placing Fractions on Number Lines

For each of the problems on journal page 86, justify your conclusions by representing the approximate position of each fraction on the number line.

SRB
135

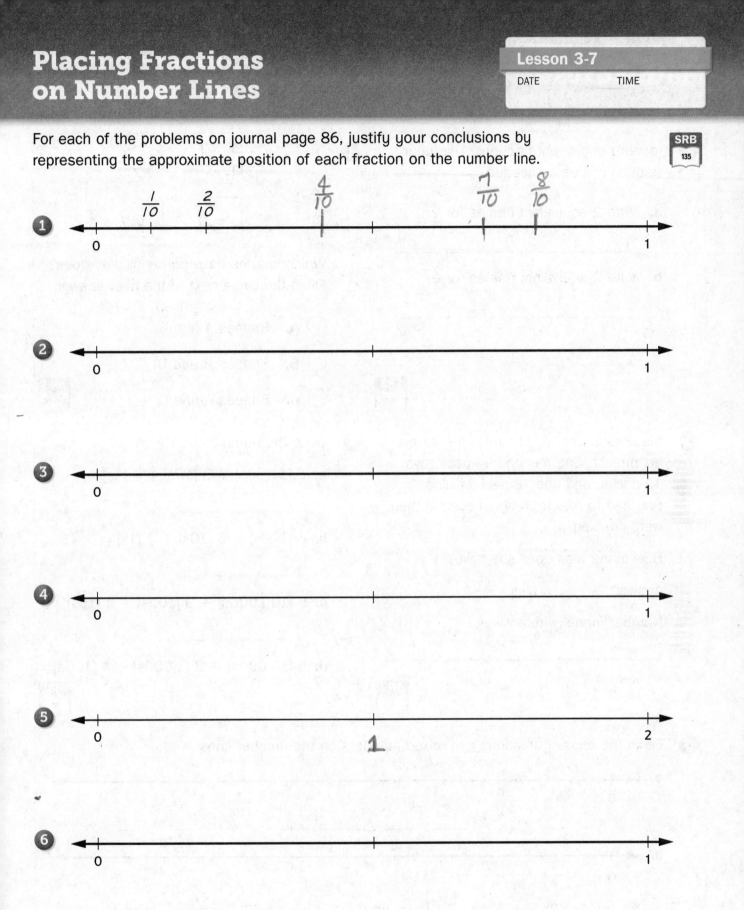

Math Boxes

Math Boxes

1 Identify equivalent fractions. Use your fraction circles as needed.

a. Write 2 equivalent names for $\frac{2}{4}$.

_____

b. Write 2 equivalent names for $\frac{2}{3}$.

_____

SRB
137

2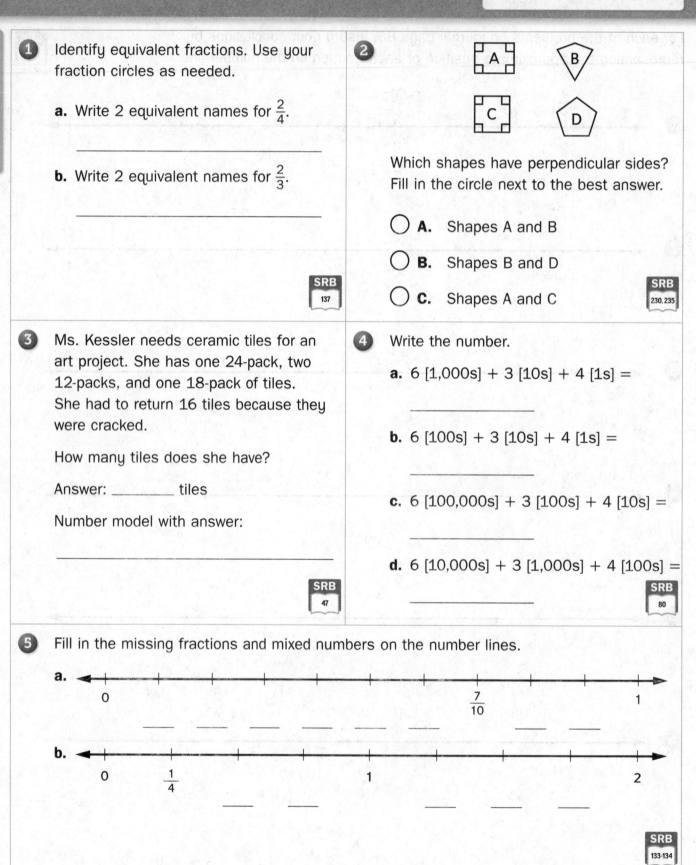

Which shapes have perpendicular sides? Fill in the circle next to the best answer.

○ **A.** Shapes A and B

○ **B.** Shapes B and D

○ **C.** Shapes A and C

SRB
230, 235

3 Ms. Kessler needs ceramic tiles for an art project. She has one 24-pack, two 12-packs, and one 18-pack of tiles. She had to return 16 tiles because they were cracked.

How many tiles does she have?

Answer: _____ tiles

Number model with answer:

_____

SRB
47

4 Write the number.

a. 6 [1,000s] + 3 [10s] + 4 [1s] =

_____

b. 6 [100s] + 3 [10s] + 4 [1s] =

_____

c. 6 [100,000s] + 3 [100s] + 4 [10s] =

_____

d. 6 [10,000s] + 3 [1,000s] + 4 [100s] =

_____

SRB
80

5 Fill in the missing fractions and mixed numbers on the number lines.

a.

0 $\frac{7}{10}$ 1

b.

0        $\frac{1}{4}$ 1   2

SRB
133-134

88

Tenths

Write a fraction and a decimal below each circle.

Whole

circle

SRB
151, 154

1

fraction: $\frac{8}{10}$

decimal: 0.8

2

fraction: $\frac{5}{10}$

decimal: 0.5

3

fraction: $\frac{3}{5} \begin{array}{l} \times 2 = \\ \times 2 = \end{array} \frac{6}{10}$

decimal: 0.6

Color part of each circle to show the decimal. Then name the fraction that is colored.

4 0.1

fraction: $\frac{1}{10}$

5 0.9

fraction: $\frac{9}{10}$

6 0.4

$\frac{4 \div 2}{10 \div 2} = \frac{2}{5}$

fraction: $\frac{4}{10}$ or $\frac{2}{5}$

Write >, =, or < to make a true number sentence.

7 0.2 $\overset{.20}{\underline{\quad < \quad}} \overset{.90}{0.9}$

8 0.7 $\underline{\quad > \quad}$ 0.6

9 1.4 $\underline{\quad > \quad}$ 1.3

10 3.2 $\underline{\quad > \quad}$ 1.8

11 0.3 $\underline{\quad < \quad}$ 0.5

12 0.1 $\underline{\quad = \quad}$ 0.1

13 2.1 $\underline{\quad < \quad}$ 2.2

14 10.1 $\underline{\quad < \quad}$ 11.1

89

Math Boxes

1 Circle the obtuse angles.

SRB
229

2 Measure the length and width of your journal to the nearest half-inch. Find its perimeter.

a. Length = _____ inches

b. Width = _____ inches

c. Perimeter = _____ inches

SRB
200

3 Complete the sentences with any of these words:

ten   place   one hundred   value

a. The value of a digit depends on its

_____ in a number.

b. The value of a digit is _____
times as large as the digit in the place to its right.

c. In the number 49,982, the digit 4 is in

the ten-thousands _____.

d. In the number 49,982, the value of

the digit 9 on the left is _____
times as large as the value of the
digit 9 on the right.

SRB
78-79

4 Kavi needs 196 lemon bars for a party. He baked 60 in the first batch and 72 in the second batch. However, he threw away 15 bars because they were burned. How many lemon bars does Kavi still need to bake?

Answer: _____ bars

Number model with answer:

_____

SRB
47, 83

5 **Writing/Reasoning** Use words from Problem 3 to compare the 8 in 8,000 and 80,000.

SRB
78-79

_____

_____

_____

90

Tenths and Hundredths

Write the fraction and equivalent decimal for each grid.

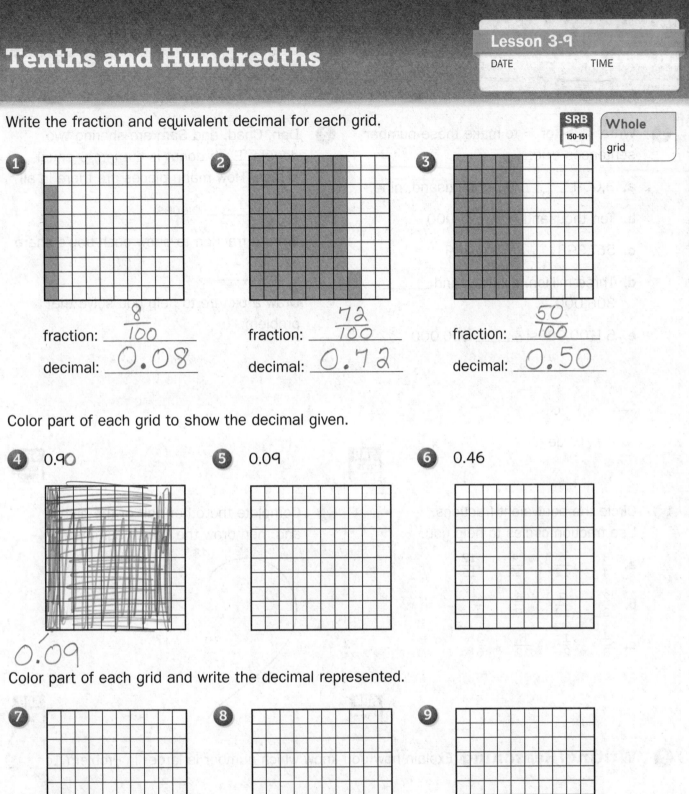

1

fraction: $\dfrac{8}{100}$

decimal: 0.08

2

fraction: $\dfrac{72}{100}$

decimal: 0.72

3

fraction: $\dfrac{50}{100}$

decimal: 0.50

SRB
150-151

Whole

grid

Color part of each grid to show the decimal given.

4 0.90

0.09

5 0.09

6 0.46

Color part of each grid and write the decimal represented.

7

decimal: _____

8

decimal: _____

9

decimal: _____

Math Boxes

1 Write <, >, or = to make these number sentences true.

a. 3,009 _____ Three thousand, nine

b. Ten thousand _____ 1,000

c. 567,398 _____ 567,489

d. Three million, six thousand _____ 306,000

e. 5 [100,000s] _____ 500,000

SRB
78-79,
81

2 Dan, Chad, and Sam are sharing two donuts. Each donut is cut into 6 equal pieces. How many pieces are there in all?

_____ pieces

Write a fraction to show each boy's share.

_____

Draw a picture to help you solve the problem.

SRB
125-126,
156-157

3 Circle the equivalent fractions. Use fraction circles to help you.

a. $\frac{3}{4}$ $\frac{6}{12}$ $\frac{9}{12}$ $\frac{12}{16}$

b. $\frac{2}{3}$ $\frac{9}{12}$ $\frac{6}{9}$ $\frac{4}{9}$

c. $\frac{4}{8}$ $\frac{1}{2}$ $\frac{8}{12}$ $\frac{3}{6}$

SRB
137

4 Complete the other half of the picture and then draw the line of symmetry.

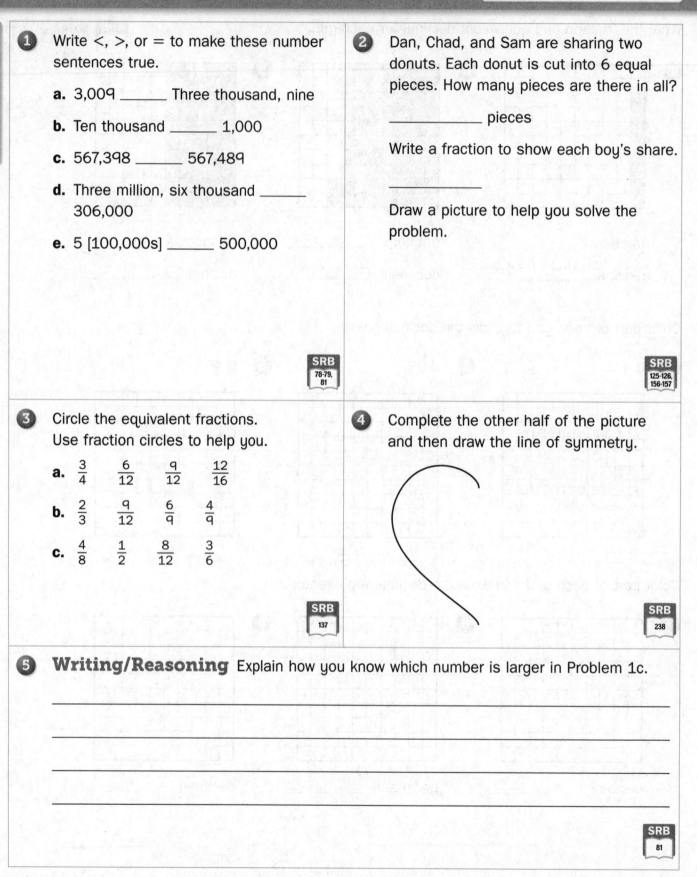

SRB
238

5 **Writing/Reasoning** Explain how you know which number is larger in Problem 1c.

_____

_____

_____

_____

SRB
81

Exploring Decimals

Work with a partner. Pick up a handful of cubes and place them on the square grid on *Math Masters,* page TA29. Record your work in the table below.

SRB
149-150

A	B	C	D
_____ hundredths	_____ , _____	0._____	
_____ hundredths	_____ , _____	0._____	
_____ hundredths	_____ , _____	0._____	
_____ hundredths	_____ tenths, _____ hundredths	0._____	
_____ hundredths	_____ tenths, _____ hundredths	0._____	
_____ hundredths	_____ tenths, _____ hundredths	0._____	
_____ hundredths	_____ tenths, _____ hundredths	0._____	
_____ hundredths	_____ tenths, _____ hundredths	0._____	
_____ hundredths	_____ tenths, _____ hundredths	0._____	
_____ hundredths	_____ tenths, _____ hundredths	0._____	
_____ hundredths	_____ tenths, _____ hundredths	0._____	
_____ hundredths	_____ tenths, _____ hundredths	0._____	
_____ hundredths	_____ tenths, _____ hundredths	0._____	

Math Boxes

1 a. Kim had the following packages of rubber bands: five 10-packs, three 1,000-packs, and eight 100-packs. Write the total number of rubber bands in expanded form.

_____

b. Cary has these boxes of nails: thirty 100-nail boxes, ten 1,000-nail boxes, and four 10-nail boxes. Write the total number of nails in expanded form.

_____

SRB
80

2 John's rectangular backyard has an area of 56 square feet. The width of his backyard measures 8 feet. What is the length of his backyard?

Answer: _____ feet

SRB
204

3 Multiply.

a. 9 8
   * 4
   ____

b. 6 3
   * 8
   ____

SRB
103-106

4 Roofing shingles come in bundles of 26 pieces. A roofer bought 50 bundles to cover a shed. She used 1,156 shingles to complete the new roof. How many shingles are left?

Answer: _____ shingles

Number model with answer:

_____

SRB
47

5 Solve.

a. $5 * 9 =$ _____; $500 * 90 =$ _____; $9,000 * 50 =$ _____

b. $8 * 7 =$ _____; $70 * 80 =$ _____; $800 * 700 =$ _____

c. $9 * 7 =$ _____; $900 * 7 =$ _____; $700 * 9,000 =$ _____

d. $7 * 7 =$ _____; $700 * 70 =$ _____; $7,000 * 7 =$ _____

SRB
42-46

94

SRB
149-150,
154

1 Fill in the missing information. Put longs and cubes end to end on a meterstick
to help you.

Length in Centimeters	Number of Longs	Number of Cubes	Length in Meters
36 cm	3	6	0.36 m
0.3 cm	0	3	0.03 m
8 cm	0	8	0.08 m
30 cm	3	0	0.3 m
43 cm	4	3	0.43 m
168 cm	16	8	1.68 m

Work with a partner. Each partner uses base-10 blocks to make one length
in each pair. Compare the lengths and use <, >, and = to record results.

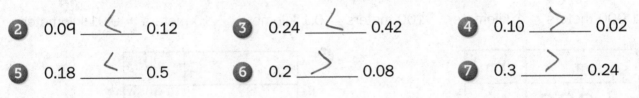

2 0.09 ___<___ 0.12   **3** 0.24 ___<___ 0.42   **4** 0.10 ___>___ 0.02

5 0.18 ___<___ 0.5   **6** 0.2 ___>___ 0.08   **7** 0.3 ___>___ 0.24

8 Follow these directions on the ruler below. Use base-10 blocks to help you.

  a. Make a dot at 4 cm and label it with the letter A.

  b. Make a dot at 0.1 m and label it with the letter B.

  c. Make a dot at 0.15 m and label it with the letter C.

  d. Make a dot at 0.08 m and label it with the letter D.

Measurement Equivalents

Complete the 2-column tables for centimeter and meter equivalencies.

SRB
182-183

100 centimeters = 1 meter   10 centimeters = 0.1 meter   1 centimeter = 0.01 meter

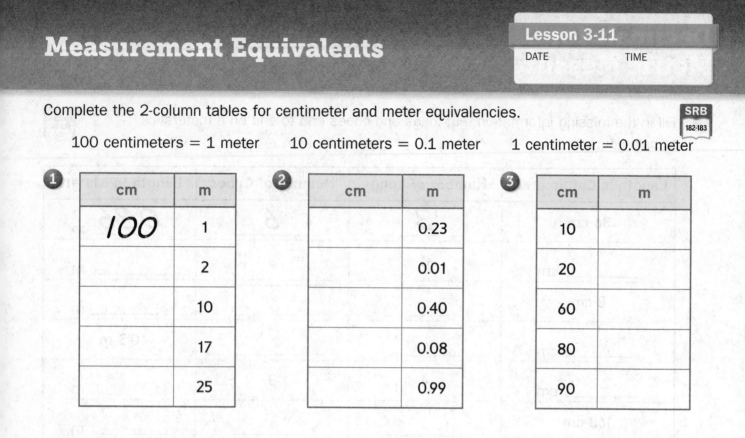

1

cm	m
100	1
	2
	10
	17
	25

2

cm	m
	0.23
	0.01
	0.40
	0.08
	0.99

3

cm	m
10	
20	
60	
80	
90	

Try This

Complete the 2-column tables for meter and kilometer equivalencies.

1,000 meters = 1 kilometer   100 meters = 0.1 kilometer   10 meters = 0.01 kilometer

4

m	km
1,000	1
	3
	7
	10
	40

5

m	km
	0.2
	0.4
	0.80
	0.05
	0.13

Math Boxes

Math Boxes

1 In the number 724,191, the digit 1 appears twice. Compare the values of the digits by completing this sentence:

The digit 1 to the left is _____ times as large as the digit 1 to the right.

SRB
78-79

2

Shade more than 0.3 but less than 0.5 of the grid.

How many boxes did you shade?

_____

SRB
150

3 Jodi's chapter book is 6 times as long as her little sister's picture book, which has 8 pages. How many pages is Jodi's chapter book?

Equation with unknown:

_____

Answer: _____ pages

SRB
56-57

4 Use U.S. traditional subtraction.

a.   7, 2 4 5
    − 4, 8 3 9

b.   4 3, 0 0 0
    − 1 2, 5 7 8

SRB
100-101

5 **Writing/Reasoning** In Problem 4b there are zeros in the top number. What steps did you take to subtract?

_____

_____

SRB
100-101

97

Math Boxes

Centimeters and Millimeters

1 Cut out the ruler given to you by your teacher. Use it to measure the pencils to the nearest centimeter.

SRB
180

a.

118mm

Pencil A is about ___*12*___ cm long.

b.

123mm

Pencil B is about ___*12*___ cm long.

2 One pencil is longer than the other. Which pencil is longer? Circle your answer.

Pencil A   (Pencil B)

3 How did you figure out which pencil is longer?

_____

_____

_____

4 Marco wants to know the difference in length between the two pencils. Can you tell him? Why or why not?

Yes! 5mm
The pencils are different
sizes when you measure
exactly.

5 Use the centimeter/millimeter ruler to measure the pencils to the nearest tenth of a centimeter and millimeter.

a. Pencil A is ___*11.8*___ cm long. Pencil A is ___*118*___ mm long.

b. Pencil B is ___*12.3*___ cm long. Pencil B is ___*123*___ mm long.

c. Pencil B is ___*0.5*___ cm or ___*5*___ mm longer than Pencil A.

98

How Wet? How Dry?

```
22 ─┬─ cm
21 ─┤
20 ─┤
19 ─┤
18 ─┤
17 ─┤
16 ─┤
15 ─┤
14 ─┤
13 ─┤
12 ─┤
11 ─┤
10 ─┤
 9 ─┤• Topeka
 8 ─┤
 7 ─┤
 6 ─┤
 5 ─┤
 4 ─┤
 3 ─┤
 2 ─┤
 1 ─┤
 0 ─┴
```

1. Use the scale at the left and the map on page 154 of the *Student Reference Book* to draw a dot to show the amount of precipitation for each of the following cities: Los Alamos, Chicago, Oklahoma City, and Mobile. Write the name of the city next to the dot.

2. Which part of the country has the most rainfall in September?

   _____

3. Which parts of the country have the least?

   _____ and _____

4. Which city gets about 5 centimeters less precipitation than Mobile?

   _____

5. Which city gets about half as much precipitation as Oklahoma City?

   _____

6. Which city gets about 3 times as much precipitation as Salt Lake City?

   _____

7. Which city gets about 2 times as much precipitation as Oklahoma City?

   _____

Reviewing Quadrilaterals and Right Triangles

SRB
233-235

1 Circle the polygon(s) that have 2 sets of parallel sides.

2 Circle the triangles that appear to be right triangle(s).

3 Circle the shape(s) without parallel sides.

4 Circle the shape(s) with perpendicular sides.

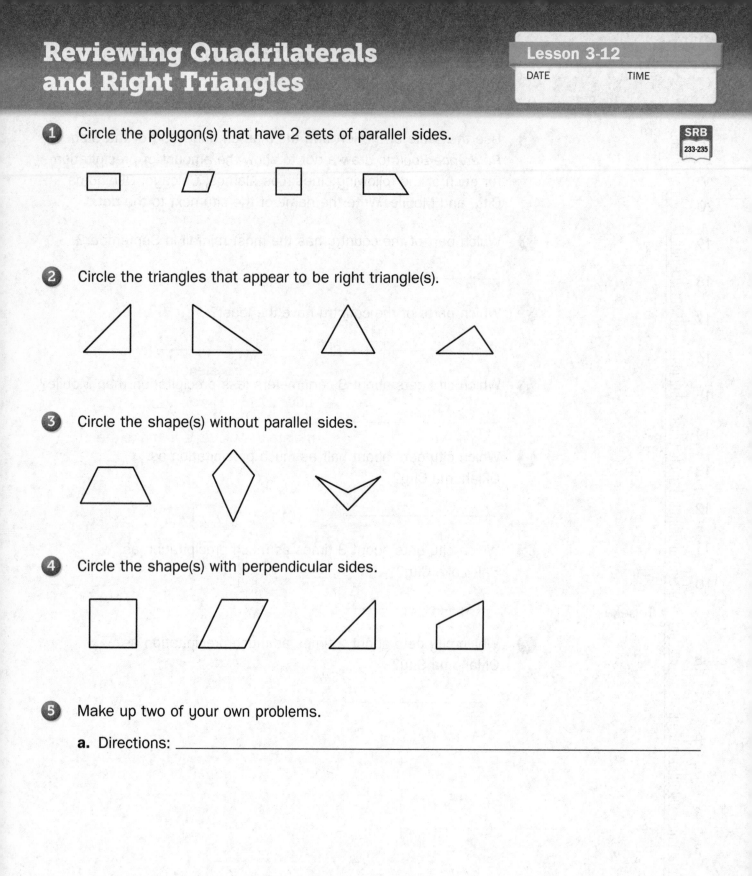

5 Make up two of your own problems.

   a. Directions: _____

   b. Directions: _____

Math Boxes

1 Write *T* for true or *F* for false.

a. _____ sixteen hundred = 1 [1,000s] + 6 [100s]

b. _____ 12 [100s] = 120,000

c. _____ 4,306 > 4 [1,000s] + 3 [10s] + 6 [1s]

d. _____ 7 thousand, four < 7,400

e. _____ 144 + 377 = five hundred twelve

SRB
78-79,
81

2 a. Divide the three graham crackers to show equal portions for 4 girls. Show one girl's share.

b. What fraction of a single cracker did each girl get?

_____ cracker

c. Write an equivalent fraction for your answer above. _____

SRB
125-126,
156-157

3 Using your fraction circles, find the fraction that is equivalent to $\frac{4}{5}$. Choose the best answer.

$\frac{9}{10}$

$\frac{8}{10}$

$\frac{8}{9}$

$\frac{6}{7}$

SRB
137

4 Complete the pictures to make them symmetrical.

SRB
238

5 **Writing/Reasoning** How did you know which fraction was equivalent to $\frac{4}{5}$ in Problem 3?

_____

_____

_____

_____

_____

SRB
137

Comparing Decimals

1 Color each grid to show the decimal. Circle the grid that shows the larger decimal.

Unit
Grid

SRB
154-155

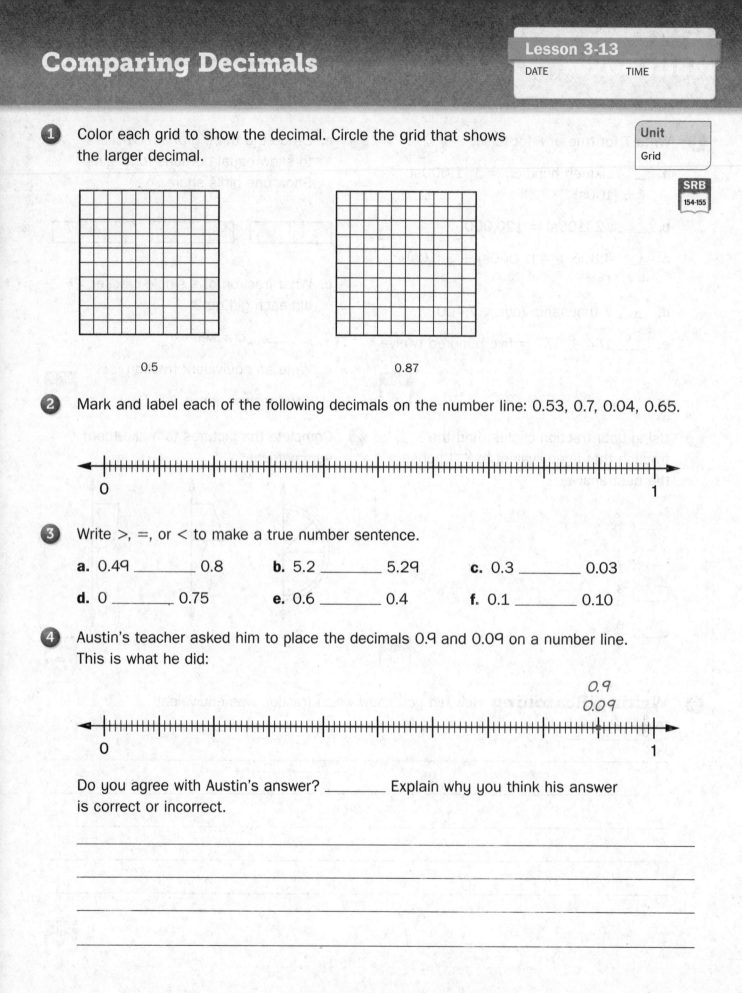

0.5   0.87

2 Mark and label each of the following decimals on the number line: 0.53, 0.7, 0.04, 0.65.

0   1

3 Write >, =, or < to make a true number sentence.

a. 0.49 _____ 0.8   **b.** 5.2 _____ 5.29   **c.** 0.3 _____ 0.03

d. 0 _____ 0.75   **e.** 0.6 _____ 0.4   **f.** 0.1 _____ 0.10

4 Austin's teacher asked him to place the decimals 0.9 and 0.09 on a number line. This is what he did:

0.9
0.09

0   1

Do you agree with Austin's answer? _____ Explain why you think his answer is correct or incorrect.

_____

_____

_____

_____

Math Boxes

① Compare the duplicate digits in each number by completing the sentences.

a. 5,**33**8   The digit 3 to the left is

_____ times as large as the 3 to the right.

b. 5,6**52**,813 The digit 5 to the left is

_____ times as large as the 5 to the right.

SRB
78-79

② Shade 0.72 of the grid.

I shaded $\dfrac{}{100}$ squares.

SRB
150

③ Write a multiplicative comparison number story for the equation 20 * 6 = 120.

_____

_____

_____

SRB
56-57

④ Use U.S. traditional subtraction.

a.
```
  8, 0 0 4
− 2, 0 0 5
```

b.
```
  2 5, 4 8 1
− 2 2, 9 6 2
```

SRB
100-101

⑤ **Writing/Reasoning** Explain how you used U.S. traditional subtraction to solve Problem 4b.

_____

_____

_____

_____

SRB
100-101

103

1 Write the number in standard form.

6 [100,000s] + 2 [10,000s] + 3 [1,000s]
+ 4 [100s] + 8 [10s] + 5 [1s]

_____

SRB
80

2 Ariel's rectangular floor has an area of 72 square feet. One side of her floor measures 9 feet. How long is the other side of her floor?

Answer: _____ feet

SRB
204

3 Multiply.

a. 3 7
    * 8

b. 5 5
    * 9

SRB
103-106

4 Hubert is stacking canned vegetables on shelves in a warehouse. Eight cans of corn are packed in each box. Beets are packed 6 cans to a box. If Hubert puts 56 boxes of corn and 92 boxes of beets on the shelves, how many cans of vegetables are on the shelves?

Answer: _____ cans

Number model with answer:

_____

SRB
47

5 Fill in the blanks.

a. _____ * 8 = 40   8 * _____ = 56   9 * 4 = _____

b. 40 * _____ = 160   _____ = 9 * 50   60 * 8 = _____

c. 200 * 60 = _____   _____ * 90 = 6,300   80 * 500 = _____

d. 2,000 * _____ = 10,000   40 * _____ = 20,000   _____ * 500 = 400,000

SRB
42-46

Multiplying Ones by Tens and Hundreds

Write a multiplication fact for each Fact Triangle. Then extend the fact by changing the first factor, first to a multiple of 10 and then to a multiple of 100.

SRB
102

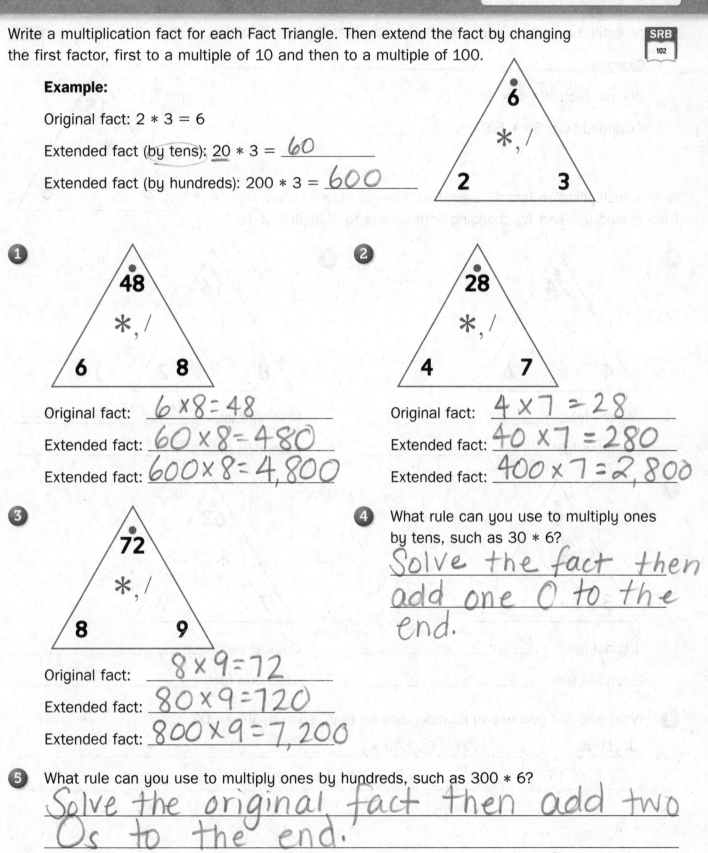

Example:

Original fact: 2 * 3 = 6

Extended fact (by tens): 20 * 3 = __60__

Extended fact (by hundreds): 200 * 3 = __600__

①

Original fact: __6 × 8 = 48__

Extended fact: __60 × 8 = 480__

Extended fact: __600 × 8 = 4,800__

②

Original fact: __4 × 7 = 28__

Extended fact: __40 × 7 = 280__

Extended fact: __400 × 7 = 2,800__

③

Original fact: __8 × 9 = 72__

Extended fact: __80 × 9 = 720__

Extended fact: __800 × 9 = 7,200__

④ What rule can you use to multiply ones by tens, such as 30 * 6?

Solve the fact then add one 0 to the end.

⑤ What rule can you use to multiply ones by hundreds, such as 300 * 6?

Solve the original fact then add two 0s to the end.

105

Multiplying Tens by Tens

You can extend a multiplication fact by making both factors multiples of 10.

Example:

Original fact: 3 * 5 = 15

Extended fact: 3**0** * 5**0** = ___1,500___

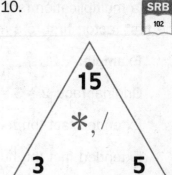

Write a multiplication fact for each Fact Triangle shown below.
Then extend this fact by changing both factors to multiples of 10.

1

Original fact: ___4×6=24___

Extended fact: ___40×60=2,400___

2

Original fact: ___8×2=16___

Extended fact: ___80×20=1,600___

3

Original fact: ___3×3=9___

Extended fact: ___30×30=900___

4

Original fact: ___9×7=63___

Extended fact: ___70×90=6300___

5 What rule can you use to multiply tens by tens, such as 40 * 60?

___Solve the original fact___
___then add 2 zeros.___

Math Boxes

1 Use your fraction circles to model the fractions below. Choose True or False.

A. $\frac{1}{5} = \frac{2}{10}$ ○ True   ○ False

B. $\frac{1}{4} = \frac{2}{6}$ ○ True   ○ False

C. $\frac{2}{8} = \frac{1}{4}$ ○ True   ○ False

D. $\frac{5}{10} = \frac{1}{3}$ ○ True   ○ False

SRB
134-136

2 Solve using U.S. traditional addition or subtraction.

a. 5,468 + 3,977 = _____

b. 6,466 − 4,715 = _____

c. 21,293 + 44,392 = _____

d. 90,532 − 43,602 = _____

SRB
92–93,
100-101

3 In the number 457,379:

a. The value of the 7 on the left is

_____.

b. The value of the 7 on the right is

_____.

c. How many times larger is the value of the 7 on the left than the value of the 7 on the right?

_____

SRB   SRB
78-79   182-183

4 Complete the table.

mm	cm	m
5,000		5
		20
		43
		9
	10,000	

5 **Writing/Reasoning** Explain how you subtracted in Problem 2d.

_____

_____

_____

_____

_____

_____

SRB
100-101

Finding Estimates and Evaluating Answers

Write an estimate and show your thinking. Use a calculator to solve the problem.
Check that your answer is reasonable based on your estimate.

SRB
82-89

1 A housefly beats its wings about 190 times per second. A wasp can beat its wings
about 400 times per second. About how many more times does a wasp beat its wings
in 1 minute compared to a housefly?

Estimate:

$$400 - 200 = 200 \times 60 = 12,000$$
$$(400 \times 60) - (190 \times 60) =$$

Answer: __12,600__ times

Is your answer reasonable? __Yes!__ How do you know? __It's__
__only 1 away from my estimate.__

2 The best cows give about 400 cups of milk every day. The best goats give about
8 cups of milk every day. About how many more cups of milk will a cow give in
1 year than a goat?

Estimate: $\underset{\text{cow}}{400 \times 400} = 160,000 \quad \underset{\text{goat}}{400 \times 8} = 3,200$
$$160,000 - 3,200 = 156,800$$

Answer: __143,080__ cups

Is your answer reasonable? __Yes!__ How do you know? __It's__
__close to the estimate__

3 A giant anteater eats about 30,000 insects every day. An aardvark eats about
50,000 insects per day. About how many insects do 1 giant anteater and 1 aardvark
eat in 1 year?

Estimate:

$$\underset{\text{Anteater}}{30,000 \times 400} = 12000000 \quad \underset{\text{Aardvark}}{50,000 * 400} = 20,000,000$$
$$\boxed{\text{Total} = 32,000,000}$$

Answer: __29,200,000__ insects

Is your answer reasonable? __Yes!__ How do you know? _____
__It's close to our estimate__

108

Practicing Multidigit Addition and Subtraction

Fill in the missing digits.

SRB
92-93,
100-101

1

```
  □ □ , □ □
    5 6 , 9 6 9
+   3 6 , 7 8 7
─────────────────
  □ 3 , □ 5 □
```

2

```
  □         □
    6 3 , 5 2 4
+ □ 4 , 9 □ □
─────────────────
  7 8 , □ 5 2
```

3 Explain how you solved Problem 2.

_____

_____

_____

_____

4

```
        □ □ □
      6 □ □ ⁰13
      7̶ 0̶ , 6̶ 1̶ 3̶
  −   4 5 , 8 4 8
  ─────────────────
      2 □ , □ 6 □
```

5

```
    □   13  □
  □ □   3̶ 1̶ □
  5̶ 2 , □ 2 1̶
− 3 3 , 7 5 4
─────────────────
  □ □ , 6 □ □
```

6 Explain how you solved Problem 5.

_____

_____

_____

_____

Math Boxes

Math Boxes

1 Write a formula for finding the perimeter of a rectangle?

_____

30 ft

15 ft

What is the perimeter of this rectangle?

_____ ft

SRB
200

2 Mrs. Drew ordered ribbon for her fabric store. She ordered 55 meters of red ribbon, 76 meters of white ribbon, and 80 meters of blue ribbon. How many meters of ribbon did Mrs. Drew order?

Answer: _____ m

How many centimeters is that?

_____ cm

SRB
182-183

3 Isabella's small pizza from Al's Pizzeria was cut into 4 equal-size pieces. Liam's small pizza from the same place was cut into 8 equal-size pieces. Isabella ate 1 piece of her pizza. Liam ate 3 pieces of his pizza. Write fractions to show how much pizza each person ate.

Isabella: _____   Liem: _____

Who ate more? _____

Explain. _____

_____

SRB
145-148

4 Each time a baseball pitcher pitches the ball over home plate, the ball travels about 20 yards. About how far will the ball have traveled from the pitcher to home plate after 9 pitches?

Answer: _____ yards

How many feet is that?

_____ feet

SRB
186-187

5 Fill in the missing fractions and mixed numbers on the number lines.

a.

0   1

___ ___ ___ ___ ___

b.

1   2

___ ___ ___ ___   ___ ___ ___ ___

SRB
133

110

Floor Tiling

Math Message

Maya wants to lay tile on a floor that is 8 feet wide by 24 feet long. The tiles
she wants to use are 1 square foot each. How many tiles will Maya need? _____ tiles

 1   Draw a picture to represent Maya's floor.

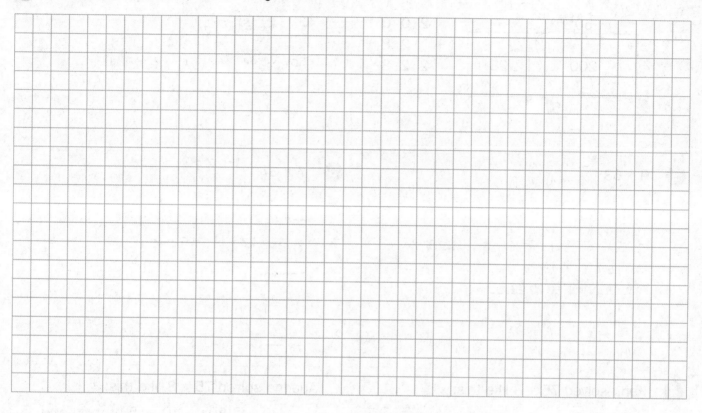

 2   Explain how you figured out how many tiles Maya needs.

_____

_____

_____

_____

Partitioning Rectangles

Use partitioned rectangles to represent each problem. Then write an equation to show how you added each part of the rectangle to get the product.

SRB
103-104

Example: $4 * 57 = 228$

	50	7
4	200	28

57

```
  2 0 0
+   2 8
  2 2 8
```

1 $5 * 48 = 240$

	40	8
5	200	40

```
  200
+  40
  240
```

2 $9 * 63 = 567$

60   3

	60	3
9	540	27

```
  540
+  27
  567
```

3 $7 * 37 = 259$

30   7

	30	7
7	210	49

4 Kadir solved 75 * 8 like this:

	70	5
8	560	40

75

```
  5 6 0
+   4 0
  6 0 0
```

Ariadne solved 75 * 8 like this:

	30	30	10	5
8	240	240	80	40

75

```
  2 4 0
  2 4 0
    8 0
+   4 0
  6 0 0
```

What is similar about their solution strategies? What is different?

_____

_____

_____

_____

Math Boxes

---

1 Write two equivalent fractions for each fraction below. Use your fraction circles, if helpful.

a. $\frac{1}{3}$ _____ , _____

b. $\frac{3}{5}$ _____ , _____

c. $\frac{2}{6}$ _____ , _____

SRB
136-137

---

2 Solve using U.S. traditional addition or subtraction.

a. $8,386 + 9,650 =$ _____

b. $1,742 - 563 =$ _____

c. $73,849 + 54,978 =$ _____

d. $38,510 - 15,496 =$ _____

SRB
92-93,
100-101

---

3 In the number 482,856, the value of the 8 on the left is

_____ .

The value of the 8 on the right is

_____ .

How many times larger is the value of the 8 on the left than the value of the 8 on the right?

_____

SRB
78-79

---

4 Which number sentence below will convert 148 meters to centimeters? Choose the best answer.

◯ $148 * 10$

◯ $148 * 100$

◯ $148 / 10$

◯ $148 + 100$

SRB
182-183

---

5 **Writing/Reasoning** Explain how you know the fractions in Problem 1a are equivalent.

_____

_____

_____

_____

SRB
136-137

113

Measuring Liquids in Metric Units

9,000 mL	9 L
8,800 mL	8.8 L
8,600 mL	8.6 L
8,400 mL	8.4 L
8,200 mL	8.2 L
8,000 mL	8 L
7,800 mL	7.8 L
7,600 mL	7.6 L
7,400 mL	7.4 L
7,200 mL	7.2 L
7,000 mL	7 L
6,800 mL	6.8 L
6,600 mL	6.6 L
6,400 mL	6.4 L
6,200 mL	6.2 L
6,000 mL	6 L
5,800 mL	5.8 L
5,600 mL	5.6 L
5,400 mL	5.4 L
5,200 mL	5.2 L
5,000 mL	5 L
4,800 mL	4.8 L
4,600 mL	4.6 L
4,400 mL	4.4 L
4,200 mL	4.2 L
4,000 mL	4 L
3,800 mL	3.8 L
3,600 mL	3.6 L
3,400 mL	3.4 L
3,200 mL	3.2 L
3,000 mL	3 L
2,800 mL	2.8 L
2,600 mL	2.6 L
2,400 mL	2.4 L
2,200 mL	2.2 L
2,000 mL	2 L
1,800 mL	1.8 L
1,600 mL	1.6 L
1,400 mL	1.4 L
1,200 mL	1.2 L
1,000 mL	1 L
800 mL	0.8 L
600 mL	0.6 L
400 mL	0.4 L
200 mL	0.2 L
0 mL	0 L

Complete the tables.

1

Liters (L)	Milliliters (mL)
1	1,000
3	
	5,000
10	
	14,000
22	

2

Liters (L)	Milliliters (mL)
1.5	1,500
4.4	
7.5	
8.8	
	9,100

3 Shade the number of liters and find the number of milliliters.

a. Shade 2.2 liters.

_____ milliliters

b. Shade 0.6 liter.

_____ milliliters

c. Shade 1.9 liters.

_____ milliliters

Solving Liquid Measurement Number Stories

Solve the problem. Complete the measurement scale and convert.

SRB
193-194

1 The hotel's kitchen uses about 2 liters of soap to wash dishes every day.
Soap is sold in 4-liter jugs. About how many milliliters of soap are left over after one day?

```
0 L          1 L          2 L✗         3 L          4 L
├────────────┼────────────┼────────────┼────────────┼──────────▶
0 mL      1,000      2,000        3,000        4,000
```

Answer: _2,000_ mL

Solve.

2 A washing machine uses about 150 liters of water per load. If the Lopez family washes
5 loads of laundry per week, about how many milliliters of water do they use per week?

$150 + 150 + 150 + 150 + 150$

$750 * 1,000$

Answer: _750,000_ mL

```
        50        100
   ┌────────┬──────────┐      500
 5 │ 250    │  500     │    + 250
   └────────┴──────────┘    ──────
        150                  750 L
```

3 Raina's fish tank holds 10 liters of water. Joel's fish tank holds 4 times that much water.
About how many milliliters of water does Joel's tank hold?

R-10L
J - 4× that much = 40L × 1,000 =

Answer: _40,000_ mL

Try This

4 A scientist measured 110 mL each of Solution A and Solution B and 190 mL each of
Solution C and Solution D. She then mixed them together. How many liters of solution
does the scientist have?

Answer: _0.600_ L

```
        2
A = 110 mL          600
B = 110 mL  ⎫            ──────
C = 190 mL  ⎬  600 mL   1,000
D = 190 mL  ⎭
──────
  600
```

115

Math Boxes

Math Boxes

1 A professional basketball court measures 94 ft by 50 ft. A high school basketball court is usually 84 ft by 50 ft. Write a formula for the perimeter of a rectangle. Use it to find the perimeter of each court.

Formula:

_____

Professional court: _____ ft

High school court: _____ ft

SRB 200

2 Three friends are making a 1-meter line with centimeter cubes. Ana has 36 cubes, Hua has 37, and Al has 46. How many extra cubes do they have? Fill in the circle next to the best answer.

◯ **A.** 119   ◯ **B.** 25

◯ **C.** 9   ◯ **D.** 19

SRB 182-183

3 Ani is baking bread. He needs $\frac{1}{4}$ cup of tapioca flour, $\frac{3}{4}$ cup of rice flour, $\frac{2}{3}$ cup of teff flour, and $\frac{1}{2}$ cup of buckwheat flour. Order the flour amounts from smallest to largest. Use a fraction tool, if needed.

_____, _____, _____, _____

Explain.

_____

_____

SRB 145-148

4 Ms. Bell is sewing 3 dresses. The ribbon requirements per dress are 78 cm, 92 cm, and 112 cm. How much ribbon does Ms. Bell need?

Answer: _____ cm

Will 3 meters of ribbon be enough for all three dresses? _____

Explain.

_____

_____

SRB 186-187

5 Fill in the missing fractions and mixed numbers on the number lines.

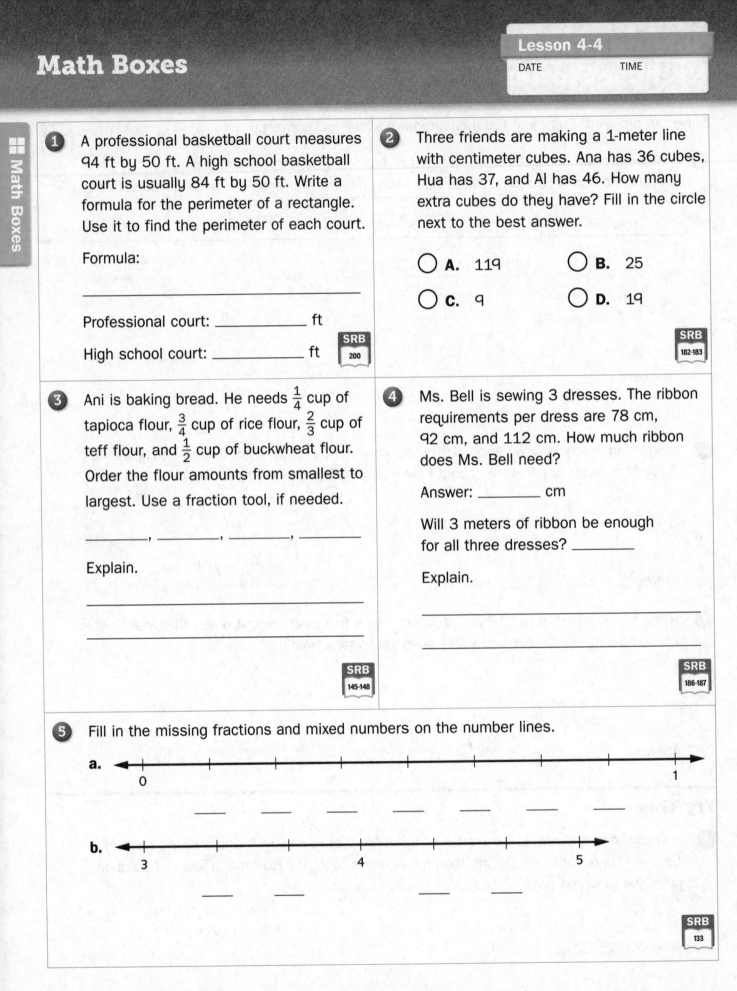

a.

0   1

b.

3   4   5

SRB 133

How Many Dollar Bills?

This picture of a dollar bill is about the same size as an actual dollar bill.
All United States bills are the same size and weight.

1 How many bills does it take to cover your book?
Make a sketch to show how to arrange the bills.

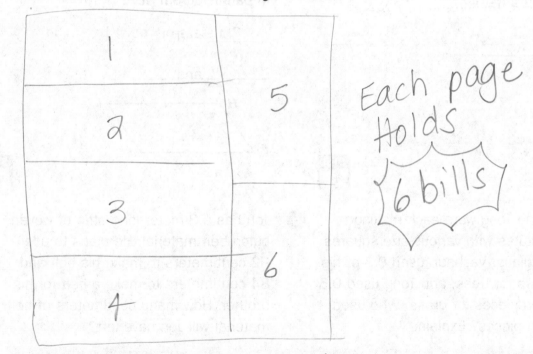

1

2   5

3

6

4

Each page
Holds
6 bills

2 Pretend the pages in your book are made of $5 bills. How much money would you have?
Show or tell how you know.

144 pages

$30 for each page

	100	40	4
3	300	120	12

300
+ 120
  12
$4,3 20

117

Math Boxes

1 Insert <, >, or = to make a true number sentence.

a. 14,357 _____ 14,275

b. 961,783 _____ 960,883

c. 656,321 _____ 665,321

d. 7,003,040 _____ 7 [millions] + 3 [1,000s] + 4 [tens]

e. Write a 7-digit number that has the digit 6 in the 10,000s place.

_____

SRB
81

2 Write a number sentence to estimate 87 * 9. Then solve. Show your work.

Estimate: _____

Answer: _____

SRB
82-89

3 Put an X in ALL the boxes that show a fraction equivalent to $\frac{2}{3}$.

☐ $\frac{3}{9}$

☐ $\frac{4}{6}$

☐ $\frac{8}{12}$

☐ $\frac{6}{9}$

☐ $\frac{6}{8}$

SRB
141

4 Name the two pairs of parallel sides in parallelogram HIJK.

_____ and _____

_____ and _____

SRB
230, 235

5 Vashaun and Tony were each making mosaic pictures with various-size squares of colored glass. Vashaun used 0.3 of his green pieces for trees, and Tony used 0.5 of his green pieces for grass. Who used more green pieces? Explain.

_____

_____

_____

SRB
125-126, 154

6 Jon has a 3-meter long strip of woven nylon belt material. He plans to use 92 centimeters to make his belt and 84 centimeters to make a belt for his brother. How many centimeters of belt material will Jon have left?

_____ centimeters

SRB
182-183

Painting Helen's Sidewalk

Math Message

Helen wants to paint the sidewalk for her block party. She needs to know the area of the sidewalk so she'll know how much paint to buy. The sidewalk is 5 feet wide and 660 feet long.

What is the area of Helen's sidewalk? _____ 3,300 _____ square feet

① Draw a picture to represent Helen's sidewalk.

600   60

5ft.   3000   300

660 ft.

② Show how you figured out the area of the sidewalk.

660 * ⑤ $600 * 5 = 3,000$
600   60 $60 * 5 = +300$
                                   $3,300 ft^2$

③ The mayor wants to beautify part of the highway by planting marigolds. She wants to plant 4 marigolds along every foot of highway for an entire mile, or 5,280 feet.

SRB
103-104,
106

How many marigolds will she need? _____ marigolds

Draw a partitioned rectangle to represent the problem. Then use partial-products multiplication to record your work in a simpler way.

Partitioned Rectangle	Partial-Products Multiplication

Exploring Partial-Products Multiplication

Draw a partitioned rectangle to represent the problem. Then use partial-products multiplication to record your work in a simpler way.

SRB
103-104,
106

1

Partitioned Rectangle	**Partial-Products Multiplication**
	$\begin{array}{r} 5\ 4 \\ *\ \ \ 7 \\ \hline \end{array}$

2

Partitioned Rectangle	**Partial-Products Multiplication**
	$\begin{array}{r} 5\ 4\ 2 \\ *\quad\ \ 7 \\ \hline \end{array}$

Exploring Partial-Products Multiplication (continued)

Use partial-products multiplication to solve the problems.

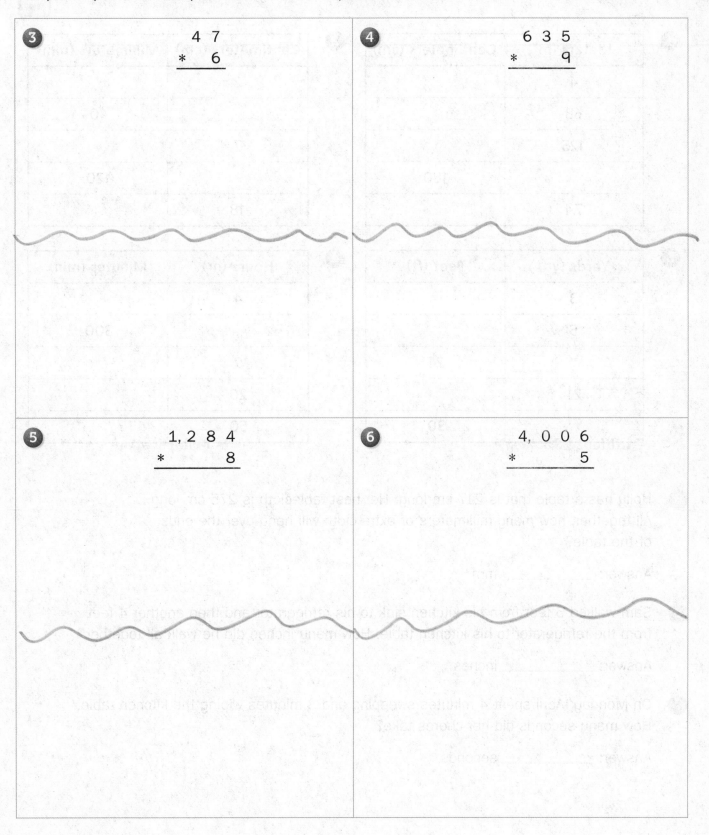

③
```
    4 7
*     6
```

④
```
  6 3 5
*     9
```

⑤
```
  1, 2 8 4
*        8
```

⑥
```
  4, 0 0 6
*        5
```

Finding Measurement Conversions

Convert.

SRB
292

1

Meters (m)	Centimeters (cm)
4	
68	
125	
	180
7.4	

2

Centimeters (cm)	Millimeters (mm)
2	
	40
7	
	120
18	

3

Yards (yd)	Feet (ft)
3	
5	
	24
21	
	30

4

Hours (hr)	Minutes (min)
4	
	300
9	
20	
50	

5 Holly has a table that is 217 cm long. Her best tablecloth is 275 cm long. All together, how many millimeters of extra cloth will hang over the ends of the table?

Answer: _____ mm

6 Sam walked 5 feet from his kitchen sink to his refrigerator and then another 4 feet from the refrigerator to his kitchen table. How many inches did he walk all together?

Answer: _____ inches

7 On Monday, Abril spent 4 minutes sweeping and 3 minutes wiping the kitchen table. How many seconds did her chores take?

Answer: _____ seconds

Math Boxes

1 A _____ is a counting number that has exactly 2 different factors.

Circle the numbers below that fit this description.

89	3	12	20	31	55
6	27	81	51	11	18
47	5	54	61	17	73

SRB
54

2 Write the formula to find the area of a rectangle.

_____

Use the formula to find the area of this rectangle.

4 cm

25 cm

Area: _____ square cm

SRB
204

3 List the factor pairs for 60.

_____

_____

_____

_____

_____

SRB
53

4 Write an equation to show each comparison.

a. Jessica is 3 times as old as her daughter Sara, who is 12. How old is Jessica?

_____

b. Ken has 60 marbles. His sister has only 12. Ken has how many times as many marbles as his sister?

_____

c. Tanya has 8 times as many books as LaToya. LaToya has 5 books. How many books does Tanya have?

_____

SRB
56-57

5 **Writing/Reasoning** Using the equation 6 * 4 = 24, write your own comparison number story similar to the ones in Problem 4.

_____

_____

SRB
56-57

123

Solving Metric Mass Problems

kg	0	5	10	15	20	25	30
g	0	5,000	10,000	15,000	20,000	25,000	30,000

SRB
188-189

Complete the two-column tables.

1

Kilograms (kg)	Grams (g)
5	5,000
9	
	12,000
18	
22	
	28,000

2

Kilograms (kg)	Grams (g)
1.5	1,500
	3,000
4.5	
6.5	
7.5	
	8,500

Solve these problems. Fill in the missing numbers on the measurement scale to help you.

3 A deer weighs 13 kilograms, and a boa constrictor weighs 17 kilograms. About how many grams heavier is the boa constrictor than the deer?

$$17 kg - 13 kg = 4 kg$$

kg	0	1	2	3	4
g	0	1,000	2,000	3,000	4,000

Answer: ___4,000___ g

4 Todd and Jorge took their dogs to the vet. Todd's dog, Duke, weighed 9 kilograms. Jorge's dog, Bruiser, weighed 2 times as much as Duke. About how many grams did Bruiser weigh?

Duke – 9kg   Bruiser – 2 x __9__ = 18kg

Answer: ___18,000___ g

124

Solving Metric Mass Problems (continued)

Solve. Use a diagram or measurement scale, if helpful.

5 Tarek bought 3 kg of hamburger, 5 kg of hot dogs, and 2 kg of steak for a barbecue. How many grams of meat did he buy?

$$3 \text{ kg}$$
$$5 \text{ kg}$$
$$\underline{2 \text{ kg}}$$
$$10 \text{ kg}$$

$$10 \text{ kg} \times 1000 = 10,000 \text{ g}$$

Answer: __10,000__ g

6 Ellen and Jim are checking in at the airport. Ellen's suitcase weighs 25,000 grams. Together, the weight of Jim and Ellen's suitcases is 49 kilograms. How many grams does Jim's suitcase weigh?

25 kg

$$\begin{array}{r} 49 \\ \text{Ellen} -25 \\ \hline \text{Jim } ? \ 24 \text{ kg} \end{array}$$

Answer: __24,000__ g

7 Buzz's Bike Shop received a shipment last week. Three bicycles each had a mass of 10 kg. Four other bikes each had a mass of 12 kg. All seven of the new bikes had baskets that each had a mass of 1 kg. How many grams total were in the shipment?

$$12 \text{ kg} \times 4 = 48 \text{ kg}$$
$$10 \text{ kg} \times 3 = 30 \text{ kg} \Big\rangle 85 \text{kg}$$
$$1 \text{ kg} \times 7 = 7 \text{ kg}$$

Answer: __85,000__ g

8 Four schools had a newspaper recycling drive last week. The following are the number of kilograms each school recycled. Decatur: 140 kg; Boone: 125 kg; Hamilton: 190 kg; Greeley: 164 kg. How many grams of newspaper were recycled in all?

$$\begin{array}{r} 140 \\ 125 \\ 190 \\ + 164 \\ \hline \end{array} \Big\rangle 619 \text{ kg}$$

Answer: __619,000__ g

9 A scientist weighed out chemicals. She needed 550 g of sodium and 450 g of cobalt. How many kilograms of chemicals did she need in all?

$$\begin{array}{r} 550 \\ + 450 \\ \hline 1,000 \text{ g} \end{array}$$

Answer: __1__ kg

125

Math Boxes

1 Insert <, >, or = to make a true number sentence.

a. 4 [100s] + 6 [10s] + 5 [1s] _____
　　　　4 [100s] + 6 [10s] + 9 [1s]

b. 27 thousand _____ 27,000

c. 3,000 + 500 + 70 _____ 3,507

d. 800,000 _____ 8 hundred thousand

e. Write a 7-digit number that has the digit 4 in the 100,000s place.

SRB
80-81
_____

2 Write a number sentence to estimate 47 * 5. Then solve the problem. Show your work.

Estimate:

_____

Answer: _____

SRB
82-89

3 Shade 5 parts of each rectangle.

_____

Label each rectangle with a fraction.

Are these fractions equivalent? Explain.

SRB
136-137

4 Draw a parallelogram. Label the vertices so that side AB is parallel to side CD.

SRB
230, 235

5 Bonnie's cup is 0.50 full of juice. Her sister's cup is different, but her cup is also 0.50 full of juice. Do they have the same amount of juice? Explain your answer.

_____

_____

_____

SRB
125-126,
154

6 One lap in the school's Olympic-size outdoor pool is 50 meters. Ben's goal is to swim 20 laps each day this week. On Tuesday he swam only 550 meters before the weather turned stormy and he had to stop. How many centimeters was he short of his goal for the day?

Answer: _____ centimeters

SRB
182-183

126

Math Boxes

1 A _____ number has more than 2 different factors.

Circle the numbers below that fit this description.

42	25	17	23	64	31
45	2	76	29	34	53
51	5	13	22	57	67

SRB
54

2 Calvin's blank art canvas measures 27 inches by 9 inches. Which number sentence shows how many square inches he has available to paint? Fill in the circle next to the best answer.

Ⓐ $27 + 9 = 36$

Ⓑ $27 - 9 = 18$

Ⓒ $27 * 9 = 243$

Ⓓ $27 / 9 = 3$

SRB
47, 204

3 List the factor pairs for 75.

_____

_____

_____

Is 75 prime or composite?

_____

SRB
53-54

4 Write an equation for each statement below, replacing the unknown with a number.

a. 35 is 7 times as much as *c*.

_____

b. 8 times as much as *y* is 72.

_____

c. 56 is *b* times as much as 7.

_____

SRB
56-57

5 **Writing/Reasoning** Read these two statements about Vince's age.

A. Vince is 2 years older than his sister, who is 10.

B. Vince's mother, age 48, is 4 times as old as Vince.

Which statement can be represented with a multiplicative comparison equation? Explain your choice.

_____

_____

_____

SRB
56-57

127

Making Travel Plans

College students in Santa Barbara, California, completed classes and are making travel plans to go home for the summer. Use the information in the chart below to help them plan their trips. If needed, draw a diagram, such as a measurement scale, to help you solve the problems.

SRB
26

One-Way Train and Bus Fare from Santa Barbara, CA			
Destination	Approximate Travel Time	By Train	By Bus
Los Angeles	3 hours	$31	$22
San Diego	6 hours	$42	$30
San Jose	8 hours	$59	$48
Oakland	9 hours	$57	$53
Sacramento	11 hours	$65	$74

1. Colleen, Emilia, and Theresa are going home to San Diego.

   a. They buy 3 train tickets using two $100 bills. How much change should they get from the cashier?

   Answer: $_____

   b. The cashier wants to use the least number of bills when she gives the girls change and has only $10 and $1 bills. How many $10 and $1 bills could she give them?

   Answer: _____

2. Juan, Terrence, Rashad, and Adrian are going home to Los Angeles. How much more would it cost for them to buy 4 train tickets than 4 bus tickets?

   Answer: $_____

128

3 How many minutes longer will the girls' San Diego trip be than the boys' Los Angeles trip?

Answer: _____180_____ minutes

6 hrs → 360 min   360
3 hrs → 180 min   -180

4 Two students buy train tickets to Oakland. Five students buy bus tickets to Oakland. How much do the tickets cost in all?

Answer: $_____379_____

57
x 2
114
Train

53 x 5
50   3

Bus

	50	3
5	250	15

265

5 Three students buy bus tickets to Sacramento. Four students buy bus tickets to San Jose. How much more do the tickets to Sacramento cost than the tickets to San Jose?

Answer: $_____30_____

Sacramento - 74 x 3 = $222
48 x 4 = $192

70	4
3 | 210 | 12

SJ

40	8
4 | 160 | 32

6 a. The trip to Oakland is how many times as long as the trip to Los Angeles?

Answer: _____3_____ times as long

O = 9 hrs.   LA = 3 hr.

b. What is the difference between the longest and shortest travel times in minutes?

Answer: _____480_____ minutes

11 - 3 = 8 hours

7 Three students take the train to Sacramento together. If they use only $1, $10, and $100 bills to pay for all the tickets, what are two different ways they could pay for the tickets?

Answer: _____65 x 3 = $195_____

$1 - 5 $1 - 5 $1 - 15
$10 - 9 $10 - 19 $10 - 8
$100 - 1 $100 - 1

Multiplying 2-Digit Numbers by 2-Digit Numbers

Draw a partitioned rectangle to represent the multiplication problem. Then use partial-products multiplication to record your work in a simpler way.

1 20 * 34 = _____

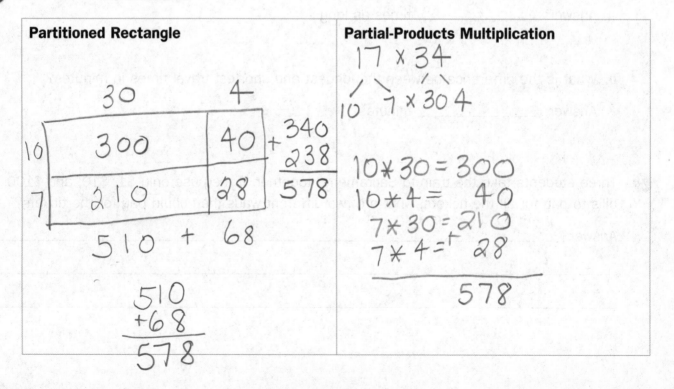

2 17 * 34 = __578__

Multiplying 2-Digit Numbers by 2-Digit Numbers (continued)

Use partial-products multiplication.

Example:

$$
\begin{array}{r}
3\ 8 \\
*\ 7\ 6 \\
\hline
2\ 1\ 0\ 0 \\
5\ 6\ 0 \\
1\ 8\ 0 \\
+\quad 4\ 8 \\
\hline
2,8\ 8\ 8
\end{array}
$$

= 30 * 70
= 8 × 70
= 6 × 30
= 6 × 8

3

$$
\begin{array}{r}
7\ 3 \\
*\ 8\ 7 \\
\hline
\end{array}
$$

4

$$
\begin{array}{r}
4\ 4 \\
*\ 2\ 8 \\
\hline
\end{array}
$$

5

$$
\begin{array}{r}
9\ 2 \\
*\ 8\ 9 \\
\hline
\end{array}
$$

6 Eli solved 28 * 37 like this.
He showed his thinking in blue:

$$
\begin{array}{r}
2\ 8 \\
*\ 3\ 7 \\
\hline
\end{array}
$$

$3 * 20 =$ 6 0
$3 * 8 =$ 2 4
$7 * 20 =$ 1 4 0
$7 * 8 = +$ 5 6

$$
\begin{array}{r}
\hline
2\ 8\ 0
\end{array}
$$

What mistake did Eli make? _____

_____

Sketch a partitioned rectangle that could help Eli understand what he did wrong.

Math Boxes

① Draw and label ray *HA*.
Draw point *T* on it.

What is another name for $\overrightarrow{HA}$? _____

SRB
226-227

② Multiply. Use the partial-products method.

_____ = 83 * 5

SRB
106

③ Round 451,062 to the nearest thousand.
Fill in the circle next to the best answer.

Ⓐ 500,000

Ⓑ 451,100

Ⓒ 452,000

Ⓓ 451,000

SRB
85-87

④ **a.** 45 = 9 * 5

45 is _____ times as many as 9

and _____ times as many as 5.

b. 7 * 8 = 56

_____ is 8 times as many as 7

and 7 times as many as _____.

c. 6 * 9 = 54

6 times as many as _____ is 54,

and 9 times as many as _____ is 54.

SRB
56-57

⑤ **Writing/Reasoning** Solve Problem 2 using a different strategy.
Show your work here and explain how this method is different.

SRB
103-104

Representing Decimals

Record the amount shaded as a decimal and as a fraction.

SRB
150-151

Whole

grid

1 0.75 $\frac{75}{100}$

2 0.40 $\frac{40}{100}$

3 0.08 $\frac{8}{100}$

Fill in the numerators to make equivalent fractions.

4 $\frac{7}{10} = \frac{70}{100}$

5 $\frac{30}{100} = \frac{3}{10}$

6 $\frac{5}{10} = \frac{50}{100}$

7 $\frac{80}{100} = \frac{8}{10}$

8 Use base-10 blocks to help you fill in the blanks.

Words	Expanded Form	Decimal	Fraction
a. 25 hundredths	2 tenths, 5 hundredths	0.25	$\frac{25}{100}$
b. 47 hundredths	4 tenths, 7 hundredths	0.47	$\frac{47}{100}$
c. 62 hundredths	6 tenths, 2 hundredths	0.62	$\frac{62}{100}$
d. 99 hundredths	9 tenths, 9 hundredths	0.99	$\frac{99}{100}$
e. 5 hundredths	0 tenths, 5 hundredths	0.05	$\frac{5}{100}$

9 seventy-seven hundredths = $\underset{\text{decimal}}{0.77}$ = $\underset{\text{fraction}}{\frac{77}{100}}$

10 three hundredths = $\underset{\text{decimal}}{0.03}$ = $\underset{\text{fraction}}{\frac{3}{100}}$

133

Math Boxes

1 Use the number line to help you add the fractions.

$$0 \quad \frac{1}{3} \quad \frac{2}{3} \quad 1$$

$$\frac{1}{3} + \frac{1}{3} = \underline{\hspace{2cm}}$$

SRB
160-161

2 Use the fraction circle to help you find the missing addend.

$$\frac{1}{4} \mid \frac{1}{4}$$
$$\frac{1}{4} \mid \frac{1}{4}$$

$$\frac{3}{4} = \frac{1}{4} + \frac{1}{4} + \underline{\hspace{2cm}}$$

SRB
160-161

3 Use a straightedge to draw the line of symmetry.

SRB
238

4 Draw ∠BAC. What is another name for ∠BAC? \underline{\hspace{2cm}}

What is the vertex of ∠BAC?

Point \underline{\hspace{1.5cm}}

C•

A•   •
                     B

SRB
228

5 Mrs. Hartman's students grew bean plants for science and measured the height of each plant. Here are their measurements in inches: 3, 4, $\frac{1}{2}$, $1\frac{1}{2}$, $1\frac{1}{2}$, 3, $2\frac{1}{2}$. Record each measurement on the line plot below.

$$0 \quad \frac{1}{2} \quad 1 \quad 1\frac{1}{2} \quad 2 \quad 2\frac{1}{2} \quad 3 \quad 3\frac{1}{2} \quad 4 \quad 4\frac{1}{2} \quad 5$$

Bean Plant Height (inches)

SRB
214

Finding the Areas of Figures

Find the area of the following shapes or objects.

SRB
104-107,
204-206

1.

25 cm

15 cm

$Area = Length * Width$
$Area = Base * Height$

Equation: __15 * 25 = a__

Answer: __375__ square centimeters

2. The science table needs to be covered with plastic for a messy experiment. The table is 25 inches wide and 38 inches long. How many square inches of plastic do we need?

Equation: _____

Answer: _____ square inches

3. Study the figure below. It is a plan for the new computer lab at Pond Cove School. The school's principal needs to determine how much carpet will be needed to cover the floor.

a. Find the area of the room. Show your work below.

4 ft   5 ft

5 ft   5 ft $25ft^2$

10 ft

5 ft
$70 ft^2$

14 ft

$5 \times \begin{array}{|c|c|} \hline 10 & 4 \\ \hline 50 & 20 \\ \hline \end{array}$

$115 ft^2$

Equations:
$(4 \times 5) + (5 \times 5) + (14 \times 5) = a$ $5 \times (4 + 5 + 14)$

Answer: __115__ square feet

b. Find the perimeter. Show your work below.

Equation:

_____

Answer: _____ feet

135

Math Boxes

1 The cupcake store earned $1,254 on Wednesday, $2,902 on Thursday, and $2,877 on Friday. On Sunday the owner counted all her money, showing that she earned $10,502 for four days. How much money did she make on Saturday?

Estimate:

_____

SRB
82-89

Answer: $_____

2 Multiply. Use the partial-products method.

_____ = 46 * 98

SRB
107

3 Write 389,457 in words.

_____

_____

_____

SRB
78-79

4 Circle ALL of the factor pairs for 56.

A.   7 and 8

B.   3 and 18

C.   2 and 28

D.   4 and 14

E.   5 and 11

SRB
53

5 **Writing/Reasoning** Was your answer reasonable in Problem 1? How do you know?

_____

_____

_____

_____

SRB
82-89

Solving Multistep Multiplication Number Stories

Write estimates and number models for each problem. Then solve.

SRB
26,
36-37

1 Danielle and Hector are selling raffle tickets to raise money for the school band. Danielle sells (5) tickets per day for 4 weeks straight. She then sells 4 tickets per day for 1 week. How many tickets does Hector need to sell if he wants to sell more tickets than Danielle?

Estimate:

$(5 \times 5 \times 5) + (4 \times 7)$

$125 + 28 = 153 \text{ tickets}$ $\boxed{154 \text{ tickets}}$

Number models with unknowns:

$(5 \times 7 \times 4) + (4 \times 7) = t$

35

Answer: More than _____ tickets

Does your answer make sense? Explain.

_____

_____

_____

2 Aliya buys 4 cartons of eggs each month. Each carton contains one dozen eggs. How many eggs does Aliya buy in two years?

Estimate:

_____

_____

Number models with unknowns:

$4 * 12 = 48$ cartons per year

$48 * 2 = 96$ cartons for 2 years $96 * 12$

Answer: $1,152$ eggs

Does your answer make sense? Explain.

	90	6
10	900	60
2	180	12

$\begin{array}{r} 900 \\ 180 \\ + 60 \\ 12 \\ \hline 1152 \end{array}$

_____

_____

137

Solving Multistep Multiplication Number Stories (continued)

Lesson 4-12

DATE   TIME

3 Antwan played goalie on both the soccer and hockey teams at school. When he played soccer, he saved 3 goals per day for 6 weeks. When he played hockey, he saved 2 goals per day for 3 weeks. How many goals did he save in all?

Estimate:

_____

_____

Number models with unknowns:

_____

_____

Answer: _____ goals

Does your answer make sense? Explain.

_____

_____

_____

4 Pablo buys 18 new songs each month. Each song costs $2. How much money does Pablo spend on songs in 3 years?

Estimate:

_____

_____

Number models with unknowns:

_____

_____

Answer: $_____

Does your answer make sense? Explain.

_____

_____

_____

138

Math Boxes

Math Boxes

1 Draw and label ray *CA*.
Draw point *R* on it.

What is another name for ray *CA*?

_____

SRB
226-227

2 Write an equation to estimate 49 * 68.
Then solve.

Estimate: _____

Answer: _____

SRB
107

3 Round to the nearest hundred-thousand.

a. 850,234 _____

b. 760,034 _____

c. 821,549 _____

d. 986,341 _____

SRB
85-87

4 Which statements correctly represent
42 = 7 * 6? Circle ALL of the correct
answers.

A. 42 is 7 times as many as 6.

B. 42 is 7 more than 6.

C. 7 is 1 less than 6.

D. 42 is 6 times as many as 7.

SRB
56-57

5 **Writing/Reasoning** Write a multiplicative comparison number story
to go along with the equation in Problem 4.

_____

_____

_____

SRB
56-57

139

Lattice Multiplication

Math Message

Column A   **Column B**

Solve using any method.

$3 * 64 = \underline{192}$

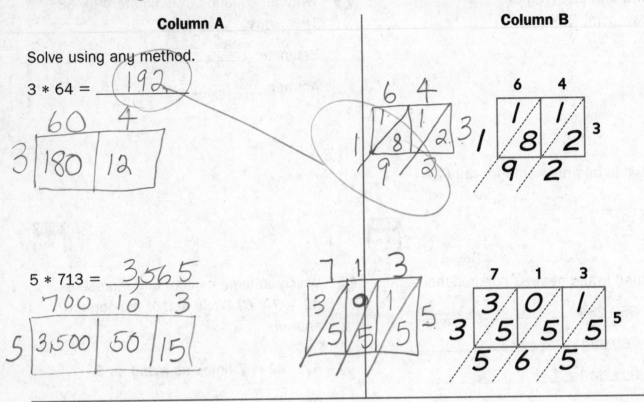

60	4
180	12

3

$5 * 713 = \underline{3,565}$

700	10	3
3,500	50	15

5

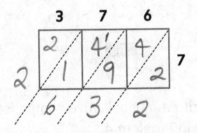

1 $7 * 376 = \underline{2,632}$

Lattice Multiplication (continued)

Use the lattice method to find the products.

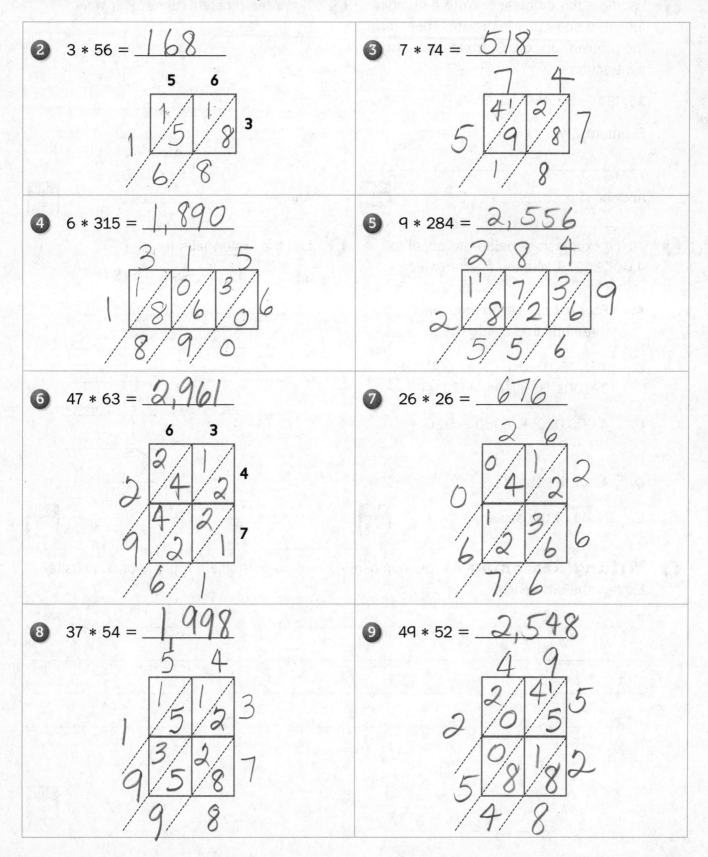

② 3 * 56 = 168

③ 7 * 74 = 518

④ 6 * 315 = 1,890

⑤ 9 * 284 = 2,556

⑥ 47 * 63 = 2,961

⑦ 26 * 26 = 676

⑧ 37 * 54 = 1,998

⑨ 49 * 52 = 2,548

Math Boxes

1 Estimate the difference. Write a number model to show your estimate. Then solve the problem using U.S. traditional subtraction.

$5,934 − $3,469

Estimate:

_____

Answer: $_____

SRB 82-89

2 Solve the problem. Show your work.

76 * 55

Answer: _____

SRB 107

3 Which expressions below are equal to 4,007,392? Circle the best answer.

A. Four million, seven thousand, three hundred ninety-two

B. 4 [1,000,000s] + 7 [1,000s] + 3 [100s] + 9 [10s] + 2 [1s]

C. 4,000,000 + 7,000 + 300 + 90 + 2

D. All of the above

E. None of the above

SRB 78-80

4 List the factor pairs for:

a. 42   **b.** 84

_____   _____

_____   _____

_____   _____

_____   _____

             _____

             _____

SRB 53

5 **Writing/Reasoning** Are the numbers 42 and 84 in Problem 4 prime or composite? Explain the difference.

_____

_____

_____

_____

SRB 54

Math Boxes
Preview for Unit 5

1 Use the number line to help you add the fractions.

$$0 \quad \frac{1}{4} \quad \frac{2}{4} \quad \frac{3}{4} \quad 1$$

$$\frac{1}{4} + \frac{2}{4} = \underline{\hspace{2cm}}$$

SRB
160-161

2 Use the fraction circle to help you find the missing addend.

$$\frac{3}{5} = \frac{1}{5} + \frac{1}{5} + \underline{\hspace{2cm}}$$

SRB
160-161

3 Using your template and a straightedge, trace a figure and draw the line of symmetry.

SRB
238

4 Draw ∠BAT. What is the vertex of ∠BAT?

Point _____

What is another name for ∠BAT?

_____

SRB
228

5 The local television station kept a record of rainfall for one week.

	Su	M	T	W	Th	F	S
Rainfall in Inches	$\frac{1}{2}$	0	$3\frac{1}{2}$	$\frac{1}{2}$	$2\frac{1}{2}$	0	0

Record each measurement on the line plot.

$$0 \quad \frac{1}{2} \quad 1 \quad 1\frac{1}{2} \quad 2 \quad 2\frac{1}{2} \quad 3 \quad 3\frac{1}{2} \quad 4$$

Rainfall in Inches

SRB
214

143

Fraction Cards 1

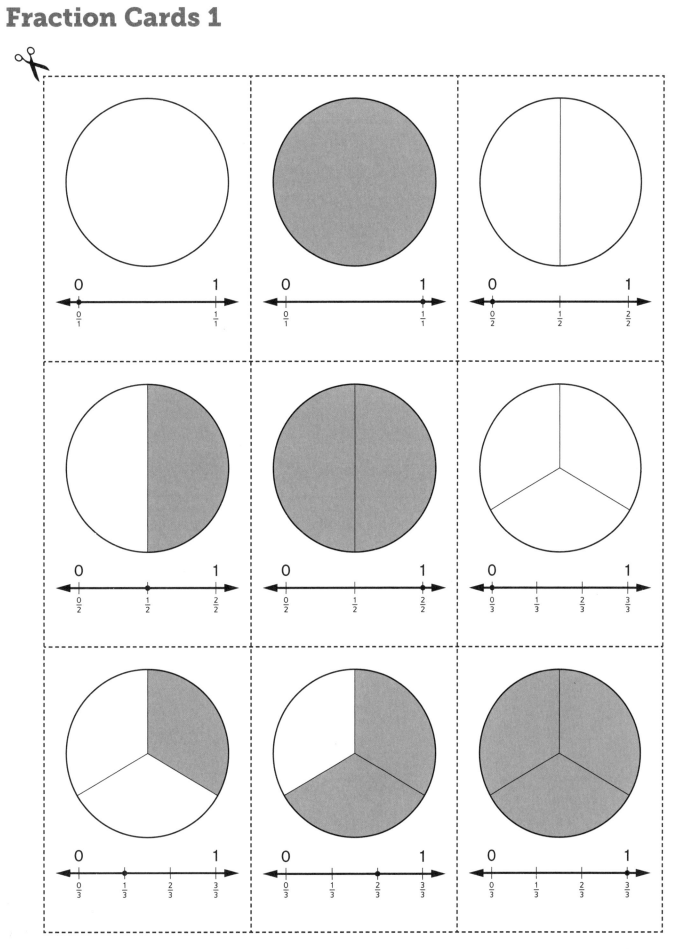

Fraction Cards 1 (continued)

$$\frac{0}{2} \qquad \frac{1}{1} \qquad \frac{0}{1}$$

$$\frac{0}{3} \qquad \frac{2}{2} \qquad \frac{1}{2}$$

$$\frac{3}{3} \qquad \frac{2}{3} \qquad \frac{1}{3}$$

Fraction Cards 2

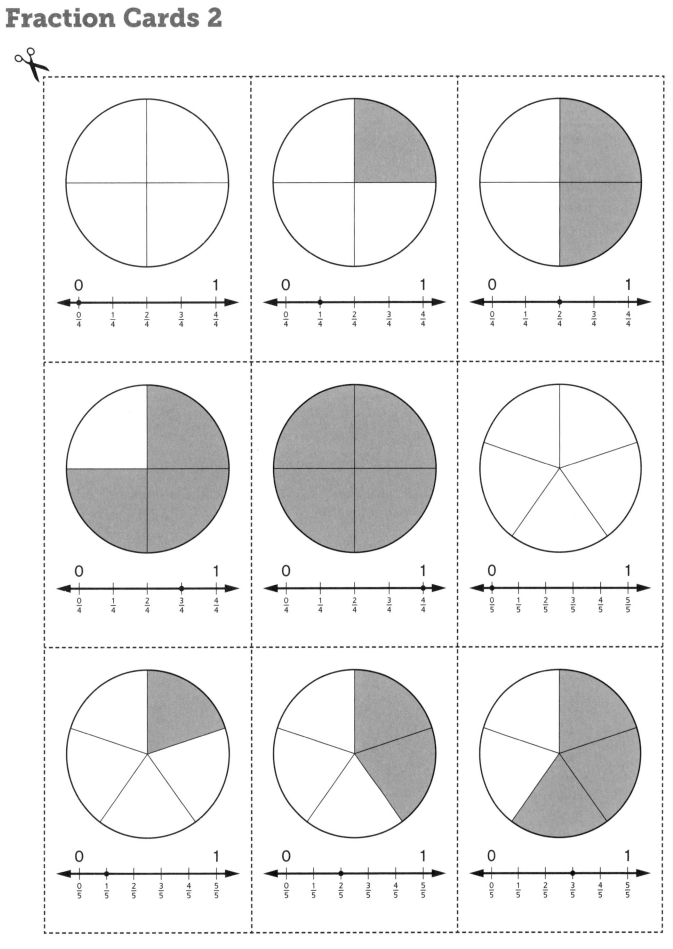

Fraction Cards 2 (continued)

$$\frac{2}{4} \qquad \frac{1}{4} \qquad \frac{0}{4}$$

$$\frac{0}{5} \qquad \frac{4}{4} \qquad \frac{3}{4}$$

$$\frac{3}{5} \qquad \frac{2}{5} \qquad \frac{1}{5}$$

Fraction Cards 3

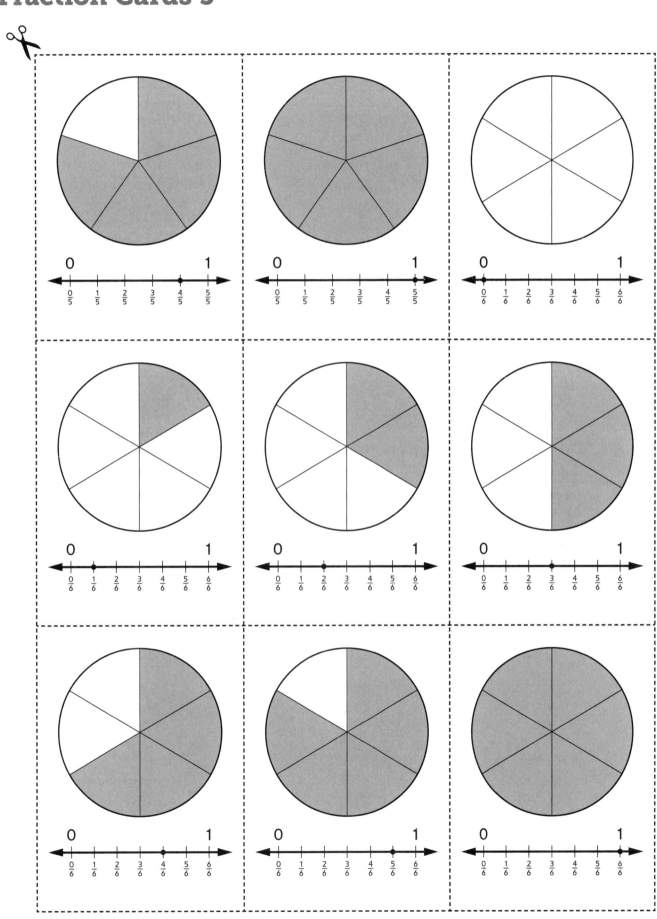

Fraction Cards 3 (continued)

$$\frac{0}{6} \qquad \frac{5}{5} \qquad \frac{4}{5}$$

$$\frac{3}{6} \qquad \frac{2}{6} \qquad \frac{1}{6}$$

$$\frac{6}{6} \qquad \frac{5}{6} \qquad \frac{4}{6}$$

Fraction Cards 4

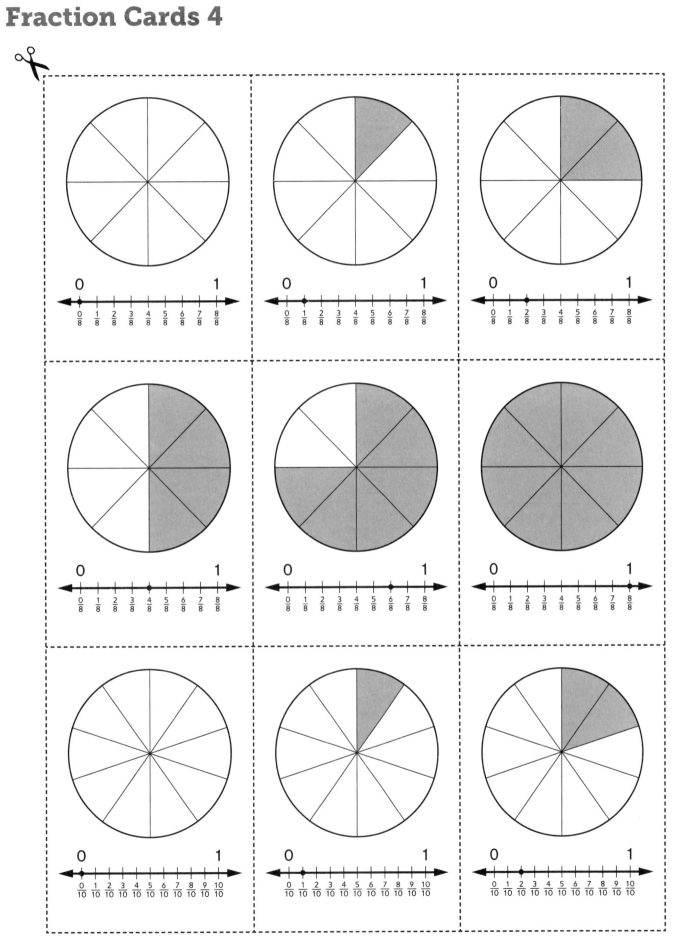

$$\frac{2}{8} \qquad \frac{1}{8} \qquad \frac{0}{8}$$

$$\frac{8}{8} \qquad \frac{6}{8} \qquad \frac{4}{8}$$

$$\frac{2}{10} \qquad \frac{1}{10} \qquad \frac{0}{10}$$

Fraction Cards 5

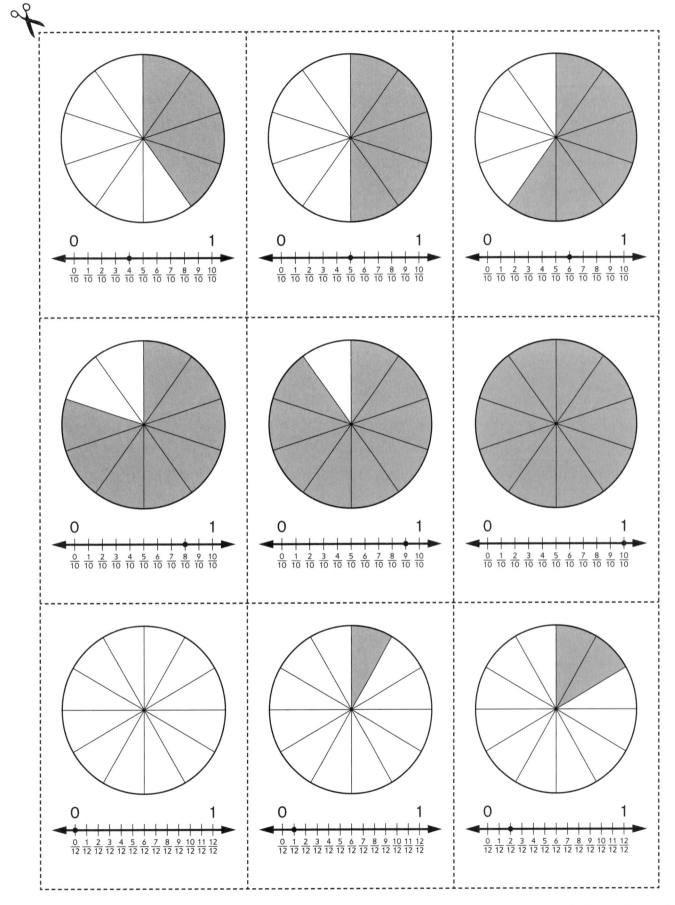

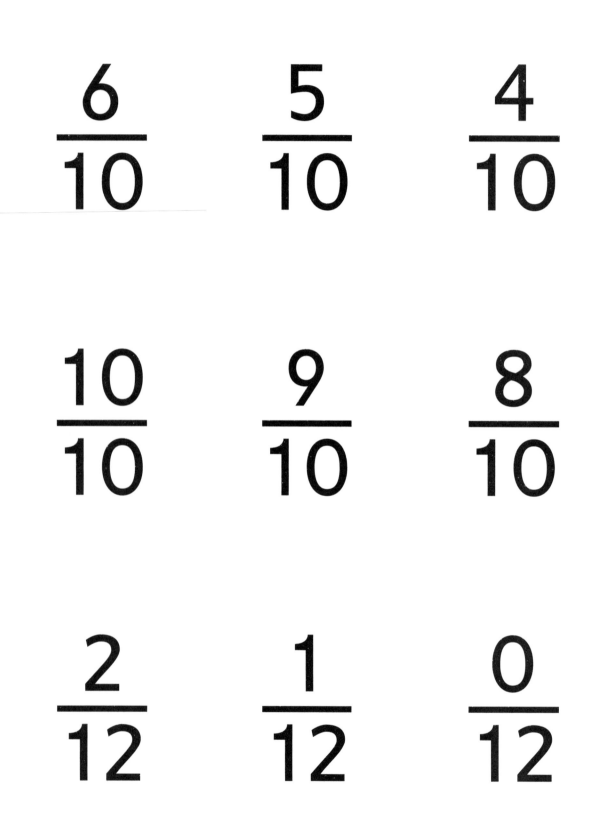

$$\frac{6}{10} \qquad \frac{5}{10} \qquad \frac{4}{10}$$

$$\frac{10}{10} \qquad \frac{9}{10} \qquad \frac{8}{10}$$

$$\frac{2}{12} \qquad \frac{1}{12} \qquad \frac{0}{12}$$

Fraction Cards 6

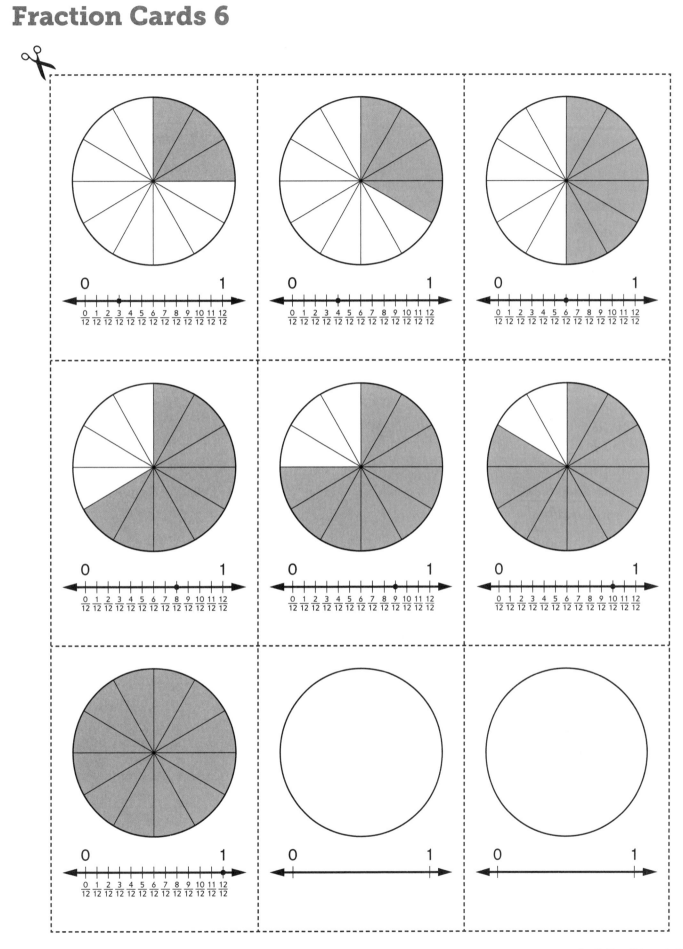

$$\frac{6}{12}$$ $$\frac{4}{12}$$ $$\frac{3}{12}$$

$$\frac{10}{12}$$ $$\frac{9}{12}$$ $$\frac{8}{12}$$

$$\frac{}{}$$ $$\frac{}{}$$ $$\frac{12}{12}$$

Fraction Circles 1

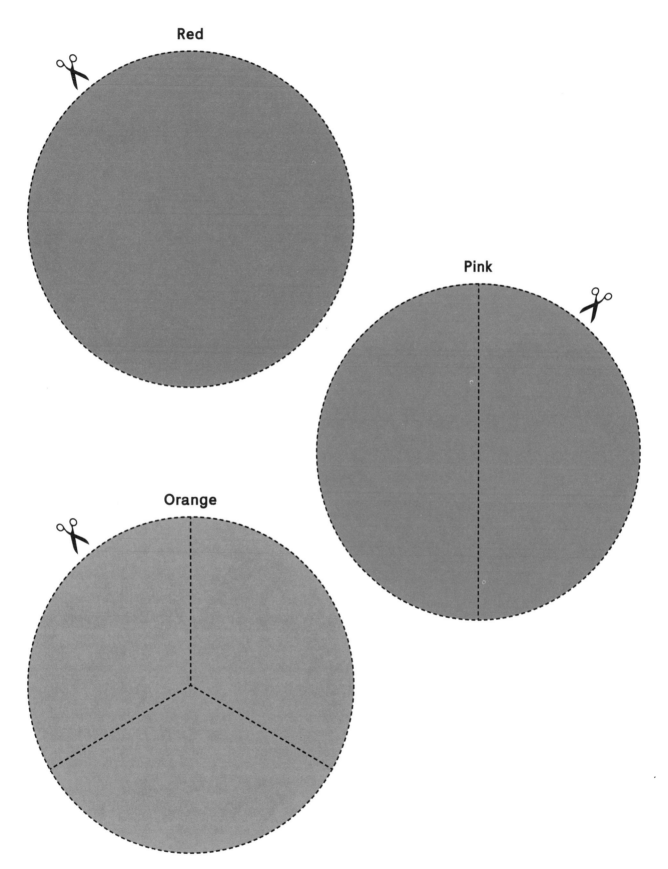

Fraction Circles 2

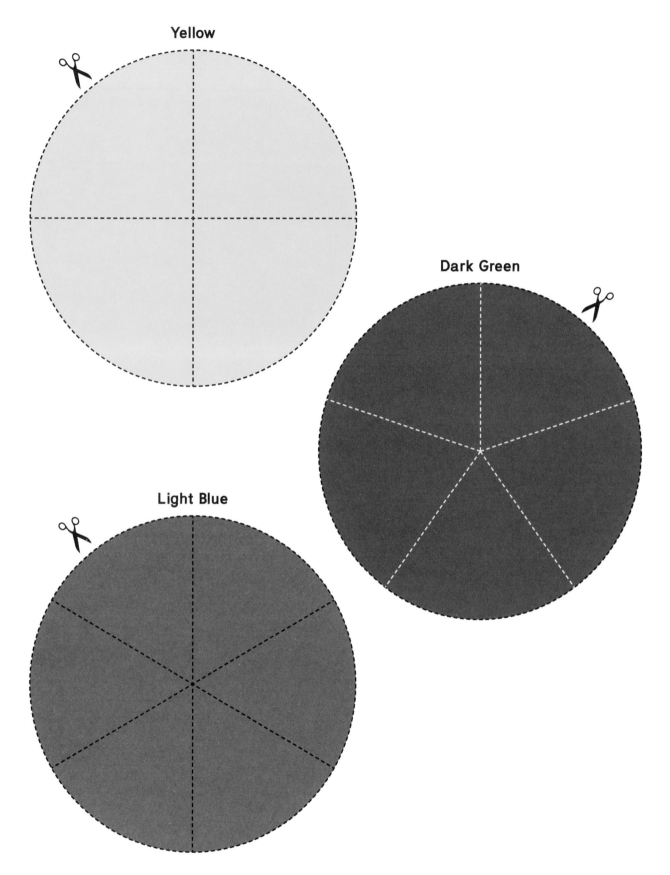

Yellow

Dark Green

Light Blue

Fraction Circles 3

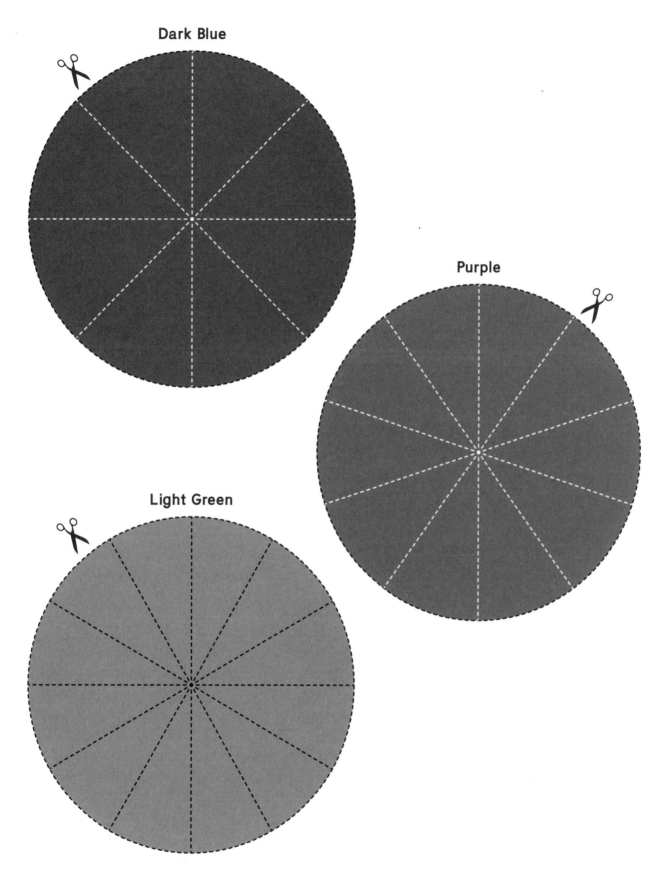

Dark Blue

Purple

Light Green

Fraction Notation Cards 1

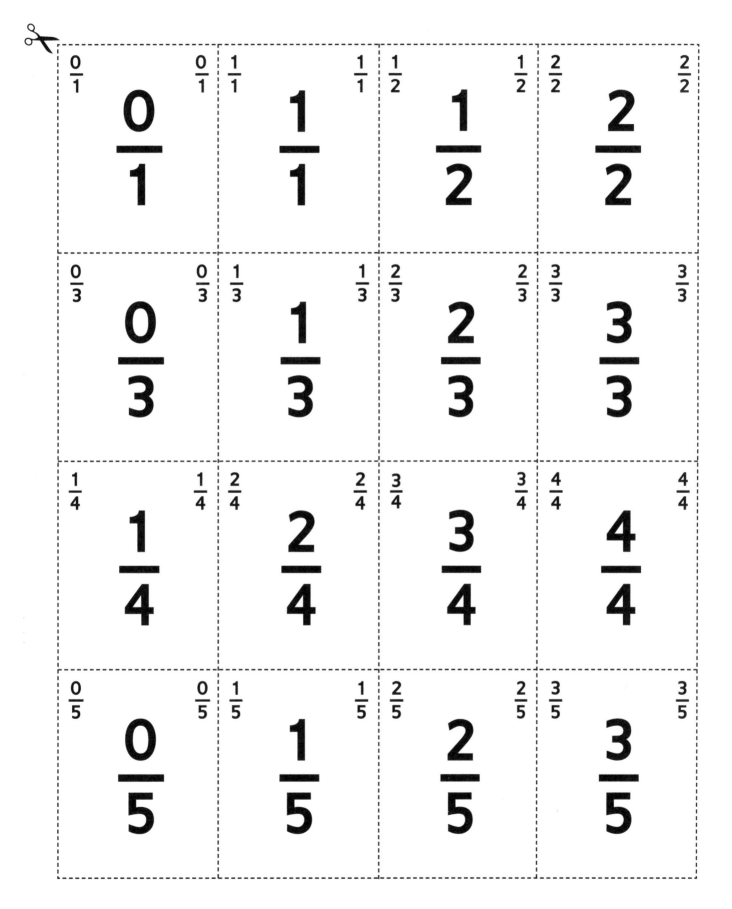

Fraction Notation Cards 2

$\frac{4}{5}$ $\frac{4}{5}$

$$\frac{4}{5}$$

$\frac{5}{5}$ $\frac{5}{5}$

$$\frac{5}{5}$$

$\frac{0}{6}$ $\frac{0}{6}$

$$\frac{0}{6}$$

$\frac{1}{6}$ $\frac{1}{6}$

$$\frac{1}{6}$$

$\frac{2}{6}$ $\frac{2}{6}$

$$\frac{2}{6}$$

$\frac{3}{6}$ $\frac{3}{6}$

$$\frac{3}{6}$$

$\frac{4}{6}$ $\frac{4}{6}$

$$\frac{4}{6}$$

$\frac{5}{6}$ $\frac{5}{6}$

$$\frac{5}{6}$$

$\frac{6}{6}$ $\frac{6}{6}$

$$\frac{6}{6}$$

$\frac{1}{8}$ $\frac{1}{8}$

$$\frac{1}{8}$$

$\frac{2}{8}$ $\frac{2}{8}$

$$\frac{2}{8}$$

$\frac{4}{8}$ $\frac{4}{8}$

$$\frac{4}{8}$$

$\frac{5}{8}$ $\frac{5}{8}$

$$\frac{5}{8}$$

$\frac{6}{8}$ $\frac{6}{8}$

$$\frac{6}{8}$$

$\frac{7}{8}$ $\frac{7}{8}$

$$\frac{7}{8}$$

$\frac{8}{8}$ $\frac{8}{8}$

$$\frac{8}{8}$$

Fraction Notation Cards 3

WILD Cards

WILD WILD **WILD** Name an equivalent fraction with a denominator of 2, 3, 4, 5, 6, 8, 10, 12, or 100.	WILD WILD **WILD** Name an equivalent fraction with a denominator of 2, 3, 4, 5, 6, 8, 10, 12, or 100.	WILD WILD **WILD** Name an equivalent fraction with a denominator of 2, 3, 4, 5, 6, 8, 10, 12, or 100.	WILD WILD **WILD** Name an equivalent fraction with a denominator of 2, 3, 4, 5, 6, 8, 10, 12, or 100.
WILD WILD **WILD** Name an equivalent fraction with a denominator of 2, 3, 4, 5, 6, 8, 10, 12, or 100.	WILD WILD **WILD** Name an equivalent fraction with a denominator of 2, 3, 4, 5, 6, 8, 10, 12, or 100.	WILD WILD **WILD** Name an equivalent fraction with a denominator of 2, 3, 4, 5, 6, 8, 10, 12, or 100.	WILD WILD **WILD** Name an equivalent fraction with a denominator of 2, 3, 4, 5, 6, 8, 10, 12, or 100.
WILD WILD **WILD** Name an equivalent fraction with a denominator of 2, 3, 4, 5, 6, 8, 10, 12, or 100.	WILD WILD **WILD** Name an equivalent fraction with a denominator of 2, 3, 4, 5, 6, 8, 10, 12, or 100.	WILD WILD **WILD** Name an equivalent fraction with a denominator of 2, 3, 4, 5, 6, 8, 10, 12, or 100.	WILD WILD **WILD** Name an equivalent fraction with a denominator of 2, 3, 4, 5, 6, 8, 10, 12, or 100.
WILD WILD **WILD** Name an equivalent fraction with a denominator of 2, 3, 4, 5, 6, 8, 10, 12, or 100.	WILD WILD **WILD** Name an equivalent fraction with a denominator of 2, 3, 4, 5, 6, 8, 10, 12, or 100.	WILD WILD **WILD** Name an equivalent fraction with a denominator of 2, 3, 4, 5, 6, 8, 10, 12, or 100.	WILD WILD **WILD** Name an equivalent fraction with a denominator of 2, 3, 4, 5, 6, 8, 10, 12, or 100.

WILD Cards (continued)

WILD WILD **WILD** Name an equivalent fraction with a denominator of 2, 3, 4, 5, 6, 8, 10, 12, or 100.	WILD WILD **WILD** Name an equivalent fraction with a denominator of 2, 3, 4, 5, 6, 8, 10, 12, or 100.	WILD WILD **WILD** Name an equivalent fraction with a denominator of 2, 3, 4, 5, 6, 8, 10, 12, or 100.	WILD WILD **WILD** Name an equivalent fraction with a denominator of 2, 3, 4, 5, 6, 8, 10, 12, or 100.
WILD WILD **WILD** Name an equivalent fraction with a denominator of 2, 3, 4, 5, 6, 8, 10, 12, or 100.	WILD WILD **WILD** Name an equivalent fraction with a denominator of 2, 3, 4, 5, 6, 8, 10, 12, or 100.	WILD WILD **WILD** Name an equivalent fraction with a denominator of 2, 3, 4, 5, 6, 8, 10, 12, or 100.	WILD WILD **WILD** Name an equivalent fraction with a denominator of 2, 3, 4, 5, 6, 8, 10, 12, or 100.
WILD WILD **WILD** Name an equivalent fraction with a denominator of 2, 3, 4, 5, 6, 8, 10, 12, or 100.	WILD WILD **WILD** Name an equivalent fraction with a denominator of 2, 3, 4, 5, 6, 8, 10, 12, or 100.	WILD WILD **WILD** Name an equivalent fraction with a denominator of 2, 3, 4, 5, 6, 8, 10, 12, or 100.	WILD WILD **WILD** Name an equivalent fraction with a denominator of 2, 3, 4, 5, 6, 8, 10, 12, or 100.
WILD WILD **WILD** Name an equivalent fraction with a denominator of 2, 3, 4, 5, 6, 8, 10, 12, or 100.	WILD WILD **WILD** Name an equivalent fraction with a denominator of 2, 3, 4, 5, 6, 8, 10, 12, or 100.	WILD WILD **WILD** Name an equivalent fraction with a denominator of 2, 3, 4, 5, 6, 8, 10, 12, or 100.	WILD WILD **WILD** Name an equivalent fraction with a denominator of 2, 3, 4, 5, 6, 8, 10, 12, or 100.

e LaunchPad activity is available for this topic. Visit **macmillanhighered.com/reinteractiveupdate.**